# ADVANCE PRAISE

"Squashed bunnies! Solar flares! Obscure Greek Island geography! The world ham radio championship puts up obstacles like no other contest, and J. K. George does a masterful job of explaining why these contestants, the Tom Sawyers of the ionosphere, are so full of good-natured obsession."

—MARK OBMASCIK, author of *The Big Year* and *Halfway to Heaven*

"*Contact Sport* should come with a warning. A casual observer of amateur radio could be inspired to become an active operator. The characters, comradery, competition, and thrill of contact come through loud and clear."

—STEWART VANDERWILT, director and general manager of KUT and KUTX Public Radio, and assistant dean of the Moody College of Communication at the University of Texas at Austin

"J. K. George has captured the essence of what makes amateur radio operators compete. The personalities of the book's characters spring to life as we meet and follow several teams and volunteers through qualification, preparation, action, and the inevitable umpiring. *Contact Sport* shines a light on the very human side of a long-enjoyed and still vital technology that is employed and enjoyed around the world. Come listen to the world turning!"

—H. WARD SILVER, author of fifteen fiction and nonfiction (technical) books and a writer for QST, the publication of the American Radio Relay League

"*Contact Sport* was thoroughly enjoyable. J. K. George did an excellent job of explaining what takes place to a nonham, while keeping it fresh for people who have been there and have been active in contesting for many years."

—JOE RUDI, former All-Star Professional Major League Baseball Player

"I enjoyed reading the details of the 2014 WRTC, and I commend J. K. George for such a great job for documenting the event."

—BOB HEIL, founder of Heil Sound, an American supplier of professional audio equipment

J.K. GEORGE

# CONTACT
# SPORT

A STORY *of* CHAMPIONS, AIRWAVES, *and a*
ONE-DAY RACE AROUND *the* WORLD

GREENLEAF
BOOK GROUP PRESS

This is a work of creative nonfiction. The events are portrayed to the best of the author's memory. While all the stories in this book are true, some names and identifying details have been changed to protect the privacy of the people involved.

Published by Greenleaf Book Group Press
Austin, Texas
www.gbgpress.com

Distributed by Greenleaf Book Group

For ordering information or special discounts for bulk purchases, please contact Greenleaf Book Group at PO Box 91869, Austin, TX 78709, 512.891.6100.

Design and composition by Greenleaf Book Group
Cover design by Greenleaf Book Group
Cover photo by Steve Moynihan (W3SM)

Cataloging-in-Publication data is available.

Print ISBN: 978-1-62634-236-1

eBook ISBN: 978-1-62634-237-8

Part of the Tree Neutral® program, which offsets the number of trees consumed in the production and printing of this book by taking proactive steps, such as planting trees in direct proportion to the number of trees used: www.treeneutral.com

TreeNeutral

Printed in the United States of America on acid-free paper

15 16 17 18 19 20   10 9 8 7 6 5 4 3 2 1

First Edition

# CONTENTS

## 1

01A—Hollis School (W1T)

## 2

02A—Kimball West (W1Z)
02B—Kimball East (W1U)

## 3

03A—Keyes Nissitissit Meadows (K1U)

## 4

04A—Heald West (N1M)
04B—Heald East (K1Z)

## 5

05A—Twin Valley North (K1W)
05B—Twin Valley East (N1T)

## 6

06B—Devens Salerno South (K1C)
06C—Devens Poplar Street (W1F)
06D—Devens Perimeter North (N1V)
06E—Devens Lake George (K1V)
06G—Devens Davao Circle (N1D)
06H—Devens Buena Vista (K1F)
06I—Devens Birch Circle (Spare)
06J—Devens Salerno North (N1Z)
06K—Devens Perimeter South (K1I)
06Q—Devens Adams Circle (W1N)

## 7

07B—Sholan South (K1P)
07C—Sholan North (N1P)

## 8

08A—State Hospital North (K1M)
08B—State Hospital East (N1A)
08C—State Hospital West (W1R)
08D—State Hospital Southwest (W1M)

## 9

09A—Wompatuck Doane SE (N1O)
09B—Wompatuck Doane Middle (W1W)
09C—Wompatuck Bunker (Spare)
09D—Wompatuck Doane NW (K1G)
09E—Wompatuck Central NW (Spare)
09F—Wompatuck Central Middle (K1O)
09G—Wompatuck Central SE (N1L)
09H—Wompatuck South Field (N1R)

## 10

10A—DevCenter Middle (W1B)
10D—DevCenter East (Spare)
10F—DevCenter South (W1D)
10G—DevCenter North (K1K)

## 11

11B—Rice West (N1G)
11C—Rice Middle (K1D)
11E—Rice East (W1S)

## 12

12A—Airport East (N1F)
12D—Airport South (K1N)

## 13

13A—Massasoit Cranberry Bog (K1R)
13B—Massasoit Water Tank (W1G)

## 14

14A—Freetown Breakneck Hill (W1L)
14B—Freetown Ranger (Spare)
14C—Freetown Dighton Rock (N1K)
14D—Freetown Sweets Knoll (W1A)

## 15

15A—Myles Standish South Cutter NW (W1C)
15B—Myles Standish South Cutter Central (W1O)
15C—Myles Standish South Cutter NE (W1K)
15D—Myles Standish South Cutter SE (W1I)
15E—Myles Standish South Cutter SW (K1S)
15F—Myles Standish North Equestrian (K1L)
15G—Myles Standish South Charge Pond D (K1T)
15H—Myles Standish North Long Pond West (N1N)
15I—Myles Standish North Long Pond East (N1U)
15J—Myles Standish South Charge Pond B (N1B)
15M—Myles Standish North Upper College
      Pond Rd (N1W)
15N—Myles Standish North Three Cornered Pond
      East (K1B)
15O—Myles Standish North Negus (W1V)
15P—Myles Standish North Kamesit (N1S)
15R—Myles Standish North 3 Corner Pond M (N1C)
15U—Myles Standish South New Grassey
      Pond East (K1A)
15W—Myles Standish South New Grassey
      Pond West (W1P)

## HQ

HQ - DoubleTree Headquarters

# ACKNOWLEDGMENTS

A nonfiction work of this sort owes much to others. There would be no 2014 World Radiosport Team Championship at all if it did not stand on the shoulders of all who came before: the previous organizers, sanctioning boards, competitors, referees, officials, volunteers, and the broad swath of amateur radio enthusiasts who provide both the audience and the general participants.

This book chronicling the July 2014 WRTC in New England came about thanks to Doug Grant, who not only offered me the opportunity to write the story, but also suggested the title, *Contact Sport*, and offered valuable critiques on several essential areas of the story. Grant's organizing skills led to a tightly run competition. With many opportunities for things to go wrong, most did not, and the exceptions were handled well by the skilled volunteers and ordinary ham radio operators who helped solve the problems and kept the event on an even keel.

Seven people read all or part of the manuscript in preliminary form and offered valuable, often extensive, critiques. This group, consisting of (in alphabetical order) Jonathan Cunitz, Rusty Epps, Doug Grant, Ward Silver, Dave Sumner, Randy Thompson, Bill Vinci, and George Wagner, pointed out a wide range of needed clarifications in depth. I can't thank them enough.

Extensive interviews via telephone with John Crovelli, Rick

Dougherty, John Laney, Steve London, Denis Pochuev, Scott Robbins, "Tree" Tyree, and George Wagner helped fill in and clarify important details. In addition to the above, email from John Barcroft, Ken Low, Dave Hawes, Sandy Räker, and Tonno Vahk added essential background. These insights enabled me to present virtual first-person descriptions in many instances, bits of information that otherwise I would not have known. Any remaining faults and inaccuracies are mine.

Several key articles in the *National Contest Journal* provided information without which this book would not have been complete. These included the entire July/August 2014 commemorative issue on the WRTC 2014, the September/October 2014 article by Rich Assarabowski on the site selection process, and the January/February 2015 issue that included the article by Carl Luetzelschwab regarding propagation and the results of the Morse code "pile-up" contests at the WRTC.

I appreciate the ability to use photographs by Bob Wilson, George Wagner, Wes Kosinski, Michael Hoeding, Manfred Wolf, Rusty Epps, Jeffrey Bail, and Steve Moynihan. Also, the high-definition signed WRTC 2014 banner photo from the hotel is courtesy of Doug Grant.

Marvin Bloomquist was kind enough to introduce me to Picasa, the photograph application, and to answer my questions. Frankly, I struggled at least as much with the photographs as I did with the writing!

It's mentioned in the book, but can't be overemphasized, that the extensive volunteer network at WRTC 2014 was amazing. No amount of kudos will be sufficient for this group of people, including Finance, Team and Referee Selection, Hospitality, Communications, Rules and Competition, the Judging Process, IT Infrastructure, Driver Support, the Beam Teams (who also installed the tents and generators), Site Selection, and the Site Management and Security volunteers who tended to so many problems and provided so much food and support to the competitors. Of course, the Organizing Committee shepherded the whole thing throughout. There would have been no WRTC, and consequently no book, without them.

I was welcomed to site 15W by Michael and Derek Bennett. They, along with Manfred Wolf, Stefan von Baltz, and Wes Kosinski, with whom I was a volunteer driver and as such more or less embedded for the weekend, provided much of the firsthand experiences. It was a pleasure to watch Wolf and von Baltz compete at a world-class level among the other elite operators.

It goes without saying that I love the hobby of amateur radio. When my Boy Scout troop visited a local ham operator in his *Nautilus*-type cocoon of mysterious glowing vacuum-tube radio gear in 1957, I was hooked. Now, many years later, the thrill is as strong as ever. I hope this book will allow some additional people, including some young people, to explore the magic of radio.

Finally, thanks to Susan Luton for editing the book before it went to the publisher. In addition, the Writers' League of Texas has provided many seminars and resources for writers, and for this one in particular. I'm pleased to be a member.

At Greenleaf Book Group, thanks to the entire team who shepherded me through the intricacies of not only a professional publishing process, but also the difficult background of release forms, since this book is nonfiction (the radio scenes do seem like magic, and they are) and actual people were involved in the story.

Last, but certainly not least, my wife, Diana, has supported this endeavor throughout. The book would not be possible without her understanding and support.

J. K. (Jim) George (N3BB)
Austin, Texas

# INTRODUCTION

Each was a highly skilled radio operator. Most had traveled thousands of miles to Boston on that sunny July weekend, ready to compete in this Olympic-style "radiosporting" event. Modern shortwave radios would be their main "weapons," but their primary asset would be the hours they'd spent honing their skills to compete at this elite level. The object of the game: to "contact"— talk to, using the human voice or the sounds of Morse code—as many people in as varied and faraway locales as possible, all within a sliver of only twenty-four hours.

It was like any other sport, really, except the playing field was invisible, comprising electromagnetic threads of energy we call radio waves. There would be winners, there would be losers, and there would be strong disagreements about the final order of finish. The two-person teams sat at stations resembling the control decks of jet planes, except they were in nylon tents in remote locations spread across eastern Massachusetts, often deep in the woods. Brilliant high-definition computer screens loomed over the radio sets, which were themselves marvels packed with vivid displays and easy-to-reach knobs and controls. The helmet in this sport was a set of headphones, with an attached, slim microphone poised and ready to make voice contact with whomever, wherever, might respond. More useful yet was the small, sleek Morse code paddle sitting to the side of each operator—allowing for even faster contacts in a language spoken the world over. Camp-style chairs, fans, assorted switches and connectors, plus maps of the world and banners of hometown radio clubs completed the radio man caves (or woman cave, in one instance).

Thousands of invisible strands of radio waves flew from each of these teams' stations out to some 10,000 amateur radio enthusiasts around the globe. With no arena, no stadium, no pumped-up fans screaming and cheering, each of the team sites remained strangely silent much of the time—like a giant video game in which the video portion is only *imagined*. (Seems strange, doesn't it? The operators envision the world at their fingertips as they make a continuous stream of rapid-fire contacts spanning many countries in different continents.) The *sounds* in this game bounce from place to place: from Boston to Germany, to Korea, then to British Columbia, and back, ricocheting off the ionosphere and landing in the ears of total strangers. Strangers who will help these contesters race to the finish.

▷ ⚬⚬⚬ ╫╟

This is a story about an international event, maybe the most intriguing one you've never heard of. In July 2014, individuals from forty countries, including fifty-nine two-person teams, sixty-three official referees, and five official judges, came together near Boston, Massachusetts, to determine the "best in the world" in a competition with absolutely no geographic boundaries in the field of play. Radio waves, after all, follow rules of physics, not man-made lines. As an intensively challenging subgenre of amateur radio (also known by the term "ham radio"), radiosporting, or ham radio contesting, is one of the most popular aspects of this globe-spanning hobby. It's fair to say that the pinnacle of this hobby is the World Radiosport Team Championship (WRTC), an Olympiad-like face-off held every four years.

Making it all the way to any WRTC involved a rigorous sequence of specific qualifying events—a major achievement in itself—and the struggle for those fifty-nine spots was intense. Of course, these radio whizzes couldn't all win—only the gold-medal team would. The rest of the field would end up from second place all the way to the final position, some with honors, others with broken dreams and bitter disappointments.

In the Olympics and most other major sporting events, elites compete against one another on fields of glory encircled by fans—the best versus the best. But this world championship was radically different. Rather than being watched by fans, the top operators of amateur radiosporting *interact* with them—with tens of thousands of enthusiasts, ranging from 10 to 100 years old, sitting at home or at a radio club with their radio setup and responding to the calls coming to them from the radiosport contestants. The more responses from these enthusiasts, the better the chance the competitors have of winning.

▷⎯ℓℓℓ⎯╟▮

This is their story. Come with me and meet a cross section of international stars of radiosporting. They represent a wide range of ages, backgrounds, and cultures. Yet each of them shares a passion for this form of magic. Follow the action as they dance upon the radio waves and see who will win and be crowned champion of their unusual sport.

WRTC competitors taking the oath of good sportsmanship. Photograph: Jim George (N3BB)

WRTC referees before taking the oath. Photograph: Jim George (N3BB)

WRTC Medallion.
Photograph: Bob
Wilson (N6TV)

*Chapter 1*

# SITE SELECTIONS

Five hundred people took a collective deep breath, and many squirmed a bit in their chairs. The constant buzz of conversations—in multiple languages—eased off, and then quieted completely.

Doug Grant, the World Radio Team Championship (WRTC) 2014 chairman, was ready to get started. "Now to a key milestone," he said as he leaned on the podium. He paused and looked out at the sea of participants crowded into the ballroom of the DoubleTree hotel in Westborough, Massachusetts. "By tradition, the defending champions will draw first."

From where I was standing at the rear of the huge room, I could see everyone crane to see the champs. At the back of the room, the Russian team, Vlad Aksenov and Alexey Mikhailov, began threading their way through the maze of tables to get to the front. They'd won gold medals in the 2010 WRTC competition in Moscow, in a nail-biter with a tough Estonian pair. (The winning margin was only 0.3 percent.) If any hard-to-read tensions from the old Soviet Union might have resurfaced four years ago, none had been reported. Aksenov reached into what appeared at a distance to be a large glass fishbowl and pulled out a fat envelope.

"You've selected site 12A," Grant announced in a crisp New

England accent. "And your referee is"—he dipped once again into the enclosure—"K7 Golf Kilo."

He didn't mention the referee's name, Denis Pochuev, since most people knew him by his radio call sign. Pochuev, a 43-year-old expert in software security now living in San Francisco, was born in Moscow and grew up in the Russian-speaking region of eastern Ukraine. He was known to this group as K7GK. (The beginning *K* indicated his current status as an American.)

The first crucial element of the WRTC, the most widely publicized and broadly followed radiosporting event in the world, had begun. The remaining fifty-eight teams would now get to draw their sites and referees. Chairman Grant called out the next team to draw: "Next up: Lima X-Ray Two Alpha and his partner, Yankee Oscar Three Japan Radio."

To most people, LX2A and YO3JR would be some sort of cryptic code. But to the men and women packed into the ballroom, these translated easily into the call signs of amateur radio stations licensed respectively by the countries of Luxembourg and Romania. Philippe Lutty, a 32-year-old civil engineer from Luxembourg with academic-looking glasses and tousled hair, led the way to the podium and selected an envelope.

"Philippe," Grant spoke slowly and waited a moment for the din to subside. "You have selected site 15P." Translated, that meant he and his teammate, Andy Ruse, a psychologist from Romania with two previous WRTC competitions under his belt, would be operating from a campsite carved out of the Myles Standish State Forest, more than seventy miles away from the headquarters hotel. "And your referee will be KC7 Victor," Grant continued.

The next team called was led by Mike Wetzel, a bearded electronics whiz and competitive tennis player in his 60s who operated almost exclusively from his brilliantly engineered home station near Indianapolis, Indiana. A group from British Columbia followed, then a Slovakian pair. A Finnish duo drew fifth.

The large room was less hushed now, with some of the tension broken. The procedure started to feel routine. As each team went through the site selection process, announcements were made, photos were taken, and the little entourage made its way to the side of the room, where an additional info package with directions and maps was provided. If needed, a driver was assigned from a pool that had been put together by an amazing volunteer network.

The ninth team to select their site and referee was the first "Special Team." Only eight of the fifty-nine teams hadn't qualified in a complex and strict meritocracy, based on numerical "points standings" from three full years of prescribed contests. Those nonqualifiers included the defending champions from the prior WRTC; two wild-card teams, invited based on some last-minute gyrations in the qualification standings; four sponsored teams; and a "Youth" team of under-25-year-olds designed to encourage younger radiosporting enthusiasts.

The Youth Team leader was a 23-year-old Italian university student who had traveled with his parents in order to compete and also to tour the United States. Filippo Vairo was the quintessential picture of a young Italian, with an outgoing personality, a shock of unruly dark hair, and free-flowing, enthusiastically accented English. Vairo had never met his teammate, a pleasant and somewhat reserved 22-year-old graduate student from Pennsylvania named Paul Whitman, until they both arrived at the hotel. The men proved to be compatible. Whitman's studies in diplomacy and international relations at Seton Hall underscored both his congenial personality and the wide diversity of his interests and talents in a highly technical hobby.

In the twelfth position, Tom Georgens and Dave Mueller represented the first of four sponsored teams. In plain lingo, "sponsored" teams bypassed the prescribed rugged qualifying procedure, and they bought their way in. The WRTC was supported by a wide range of financial contributions from many radio clubs around the world, by equipment donations and in-kind support from suppliers, and by more than four hundred volunteers—mainly from New England, but also from other regions in the United States. However, additional funding was important to meet the $750,000 level needed to conduct a first-class event. The Boston organizing committee had set the minimum level at $50,000. Only four teams had been allowed to enter in this fashion, and the leaders (and operating partners) had been required to demonstrate world-class radiosporting acumen, in addition to providing, either directly or via sponsors, significant financial muscle.

Georgens, 54 years old and the CEO of a large data storage firm, was without a doubt an elite radiosport competitor. Having no station in his home residence, he usually operated from a spartan rural building he owned on the Caribbean island of Barbados. His teammate, Mueller, was a 42-year-old career man in the US Coast Guard who had attained chief warrant officer rank. Yet with a young family and, as he said, living "on coast guard pay," he was hardly wealthy. But his track record in American radiosporting in general, and specifically in previous WRTC competitions—a silver medal in Brazil and a fourth-place finish in Finland—made him an acknowledged international force. Clearly they were a team to be reckoned with.

By now the room had settled into a sort of rhythm, as Grant continued with the station site draws, followed by photograph taking and by friends crowding around to offer best wishes to each group. The fifteenth position in the process featured a Lithuanian team. The team leader was a 58-year-old telecommunications engineer with more than forty years' worth of serious radiosporting that included two previous WRTCs. His 38-year-old partner would be competing in his first. Their American referee, a federal judge from the southern state of Georgia, was familiar to almost everyone there. He was beyond reproach—a proven veteran of decades of topflight radiosporting.

At number 21 came an American team whose leader was a 68-year-old retired engineer for the Clear Channel broadcasting company. His teammate: an electrical engineering graduate and former professional French horn player.

The parade of international luminaries, in terms of world-class shortwave communications operating skills, continued with a team from Japan. The hobby of amateur radio has been expanding rapidly in many parts of Asia. Yet, in countries such as China and Japan, the conditions of high population density, crowded and expensive housing, and difficulty building a competitive station are daunting. Europe and North America continue to be the areas where most of the radiosporting enthusiasts are located. In fact, only nine of the fifty-nine teams overall came from outside these two continents.

The random order continued with teams from Italy and England–Northern Ireland. Next, at number 28, was a twosome from Hawaii. They represented Oceania, since the state of Hawaii actually counts as a separate "country," by international convention, because of its great distance from the mainland.

The team drawing in position 29 brought about a hush in the crowd. Everyone watched closely as two Americans, 33-year-old Dan Craig and 31-year-old Chris Hurlbut, made their way to the podium. One of the favorites to win the event, this team had placed third in Moscow in 2010. In addition, Craig had won the silver medal in Brazil in 2006 and had finished a strong number 4 in 2002 in Finland, both times teaming with Dave Mueller, the coast guard officer who was paired this year on a sponsored team with business titan Tom Georgens. With Craig's record of finishing numbers four, three, and two, it didn't take a math whiz to extrapolate down to the next number. The team drew an operating site in the massive Myles Standish State Forest, located an hour and a half to the southeast, with someone very well known in radiosporting circles, the Moscow-based Igor "Harry" Booklan, as their referee.

If they were able to knock off the defending Russian champs from Moscow, Craig and Hurlbut would be doing it under the watchful eyes of an expert—and a Russian, to boot!

By coincidence, two German teams selected positions 35 and 36. The first of these included two experienced competitors: Manfred Wolf and Stefan von Baltz. Wolf, who qualified in the very competitive Central Europe region, had helped build a club station on a mountaintop in Germany; he'd selected Stefan von Baltz, a 38-year-old ophthalmologist who'd been an active amateur radio operator, both on a casual basis for informal conversations over the air and as a radiosport enthusiast, since his teenage years. Between the two, they had experience in three previous WRTCs, including operating together in the 2000 Slovenian WRTC. Von Baltz had competed in the 1996 event in the San Francisco Bay Area, so he was a veteran of nearly twenty years of elite international competition. This was a team many felt would contend for a top spot, and the crowd paid attention as their site in a remote portion of a Massachusetts state forest was announced. Their referee was "Wes" Kosinski, an electronics vocational professions teacher at a high school in Poland.

Next up was the first women's qualifying team ever in *any* WRTC event. Like Manfred Wolf, Alexandra "Sandy" Räker had qualified in the crowded Central Europe area. In her case, she took part mostly in team competitions, usually from large stations in Bavaria and Luxembourg. Räker was nothing if not determined. A former agent of the German version of the FBI, the 38-year-old had returned to her small hometown south of Hannover to head up a criminal investigation unit.

Räker's teammate was Irina Stieber, a 41-year-old marketing and sales professional who worked for a mainline newspaper in the German state of Saxony. She'd grown up near Dresden, located in the former East Germany, where she competed in a unique sporting event common in the old Eastern Bloc that had been allied with the Soviet Union. She became an expert in high-speed telegraphy, or HST, one of the designated sporting events, and was an authentic Morse code phenom. Although they had no prior experience in high-profile WRTC events, both women felt they had complementary skills, and they were committed to showing well against the "big guns" of the hobby.

The women drew a site on the grounds of a former state mental hospital, not far from the headquarters hotel. There was a noticeable buzz of approval in the room when their referee was announced as Rusty Epps, from the San Francisco Bay Area. The high-profile figure on the WRTC executive committee was a true inside player in amateur radio. A former corporate attorney and general counsel for a major telecommunications company, he had a low-key personality and avuncular demeanor that put people at ease. Epps and his longtime partner, Bill Vinci, were a standard fixture at all WRTC events.

At this point, thirty-six of the fifty-nine teams had made their way to the podium and taken care of site determination, the referee, the phalanx of photographers, finding drivers, and so on. A massive array of equipment—of bags and luggage, of radios, computers, and the supporting items that constituted two complex radio stations—began to dot the hotel lobby. These competitors and their entourages began loading cars and SUVs for the drive to their site.

As the room began to empty, the tingling atmosphere surrounding the start slowly changed as the procedure continued for the final twenty-three teams. Among them was another sponsored team, backed financially by George DeMontrond, a square-jawed, six-foot-three Texan who looked like he could be cast as the hero in a Western. He was a presence in the business community of Houston, where he owned multiple car dealerships and RV centers. In addition, he had been prominent in the socially delicious Houston Livestock Show and Rodeo (the city's self-described "signature event"), and served as a board member of the Houston Grand Opera. Besides these demands on his time, he was a lifelong (since age 12) amateur radio enthusiast who had planned, financed, and overseen a massive contest station on the family's old-money ranch near Hempstead, Texas. DeMontrond had come to rely on a team of antenna and tower specialists who had gotten into what was now a full-time business building and maintaining amateur radio stations nationwide, since many of the hobby's enthusiasts were getting older and had disposable income at the same time they were losing (if they ever had it) their ability to climb fifty or even one hundred feet up towers in order to install antennas and make all the connections that keep electrons flowing.

John Crovelli, a somewhat crusty and terse New Jerseyan, had gained DeMontrond's confidence for two reasons: his capability to build and maintain the world-class amateur radio station at the ranch and his status as a superstar radiosporting competitor. Crovelli operated from both the United States and the Dutch island of Aruba, where he rented a house and maintained an amateur radio contest station in the abrasive climate of the Caribbean. He had competed in four previous WRTCs. He had also tried hard to qualify from his region of New York and New Jersey but had fallen short. So, he was pleased when DeMontrond asked him to partner in the final open slot for a sponsored team. The Texan was primarily a voice operator, so he would focus on that, while Crovelli, who could do both tasks equally well, would handle Morse code.

Another team in the final third of the site draw included the most prominent public figure of all, a man who was sought out by reporters for National Public Radio, *The Wall Street Journal*, and major regional newspapers. Scott Redd, better known to the radiosporting community as K0DQ, had retired from the US Navy as a vice admiral. After a stint as CEO of a high-tech educational firm, he returned to Washington to take a key role as the first director of the US National Counterterrorism Center.

Redd's patience and calm presence masked a steely competitive spirit that had netted him eleven world championships in major specific radiosporting events, as well as induction into the CQ Contest Hall of Fame in 2008.

Redd's teammate and actual qualifier from the extremely competitive mid-Atlantic region (Maryland, Pennsylvania, and Delaware) was Ken Low, who worked as a financial advisor for high-net-worth families. Low had lived in both the United States and Europe because of his wife's work with the US State Department. He had an undergraduate degree in engineering but had worked on Wall Street, at least electronically, for thirty years. Besides his lifelong interest in amateur radio, Low was one of the foremost Americans competing in the Eastern European–centered high-speed telegraphy—also a specialty of Irina Stieber of the German team. He had led an American team in every international HST competition. Yet, since he and Redd were equally balanced in terms of voice and Morse, they had high hopes of doing well.

The competitors drawing at number 58 were a team to be reckoned with—everyone knew that. The team leader, Kevin Stockton, had qualified on the strength of a massive amateur radio station he and his father had constructed on a mountaintop in the Ozarks of northwestern Arkansas. Their huge antennas delivered strong signals into Europe—they were even louder in Japan. No one in Texas or the Southwest could beat him—no one, that is, but the guy he asked to be his teammate, Steve London. After retiring from a job as a network engineer with Cisco Systems in Colorado, London had moved with his wife to a remote area in, of all places, southwestern New Mexico. On an impressive mesa with favorable terrain sloping to both Europe and Asia, London continually produced radiosporting magic. One might think he had little chance against operators in Texas or Arkansas in trying to reach Europe; nor would it be likely he could beat operators along the East Coast. Yet, with his extraordinary location, his station-building acumen, and his almost mutant-like skills in both voice and Morse code, London had won numerous contests along with qualifying positions in past WRTC events. This time, however, he had backed off in order, he said, "to give others a clear field," and Stockton had become top dog from that part of the United States. Of course, the first thing Stockton did was to contact London and ask if he would join him for the WRTC 2014. London apparently was unable to resist the competitive juices and agreed. Stockton had competed once before in Moscow, and London had competed in five previous WRTC events. Everyone expected this team to be formidable.

By the time the Stockton–London team, along with their Ukrainian referee, left the podium, only one team remained. These contestants took the final envelope, and Doug Grant announced the information to a handful of people scattered among the tables that had been chock-full earlier.

By 10:30 Friday morning, the station draw was complete, and all fifty-nine teams were on their way to look over the sites where they would be setting up their stations for the contest, which would begin in just over twenty-two hours. They would wake up early Saturday morning and return to the sites for any final adjustments before the competition began at 8 a.m. Eastern Daylight Time, which correlated to 12 noon Coordinated Universal Time (UTC) on the prime meridian in the United Kingdom.

Chairman Doug Grant (K1DG) at the opening ceremony. Photograph: Bob Wilson (N6TV).

L to R: Youth Team: Filippo Vairo (IZ1LBG), referee Tibi Tebeica (YO9GZU), and Paul Whitman (WQ2N). Photograph: Bob Wilson (N6TV).

L to R: Dan Craig (N6MJ), referee "Harry" Booklan (RA3AUU), and Chris Hurlbut (KL9A). Photograph: Bob Wilson (N6TV).

L to R: Sandy Räker (DL1QQ), referee Rusty Epps (W6OAT), and Irina Stieber (DL8DYL). Photograph: Bob Wilson (N6TV).

L to R: Manfred Wolf (DJ5MW), referee Wes Kosinski (SP4Z), and Stefan von Baltz (DL1IAO). Photograph: Bob Wilson (N6TV).

L to R: John Crovelli (W2GD), referee Igor Syerikov (UT7QF), and George DeMontrond (NR5M). Photograph: Bob Wilson (N6TV).

L to R: John Sluymer (VE3EJ), referee "Tom" Soomets (ES5RY), and Fred Kleber (K9VV). Photograph: Bob Wilson (N6TV).

L to R: Referee Dmitry Stashuk (UT5UGR), Kevin Stockton (N5DX), and Steve London (N2IC). Photograph: Bob Wilson (N6TV).

# Chapter 2

# THE WOMEN'S TEAM

Sandy Räker and her teammate, Irina Stieber, brought "so much stuff, including spares for everything," that they were worried it wouldn't fit into the car that would drive them to their site. Half an hour after their site selection, surrounded by huge luggage containers, they waited in the hotel lobby for Tim Duffy to drive up in his big black SUV. The license plate, K3LR, indicated that one of the most-respected amateur radio operators in the United States was behind the wheel. He had gotten to know Räker over the air in radio contests, and she had reached out to him for advice on radiosporting as her interest deepened.

Duffy, a goal-driven but personable man, had wrapped up a successful career in telecommunications and the cellular phone industry. A distinguished hint of gray added to his clean-cut and steady demeanor; nothing seemed to faze him. His station at his home in western Pennsylvania was a well-engineered masterpiece. Over the past decade he had put together a formidable set of towers and antennas, all designed to optimize radiosporting. Inside, top-of-the-line Japanese radios and specialized home-built amplifiers ringed an impressive operating room. He had a rock-solid team of operators from all over the country who, year after year, returned to man his station, a goliath whose signal into Europe actually equaled the best setups on the East Coast—stations located much closer. Europe is a sweet spot for two reasons: It includes a high concentration of radiosporting enthusiasts, and it comprises many different countries, which count as additional diversity factors to increase the score.

Duffy had dedicated much of his life to this hobby. Now he was a senior executive in a company specializing in equipment for amateur radio. He had conceived of and organized "Contest University" at the largest amateur radio gathering in the world, in Dayton, Ohio. He was *the man*, and the man himself had become a key advisor to the young German women. They couldn't have found a better person.

Duffy's SUV was now packed with the massive trove of radio gear. Other contestants stopped to take photos of the famous Pennsylvania license plate. After all, K3LR was one of the world's best-known call signs. Finally, Räker and Stieber piled in, along with Rusty Epps, their referee. The large vehicle's luggage space was crammed full, and several smaller containers rode on laps and in crevices and crannies among the passengers when Duffy pulled out of the DoubleTree hotel and headed for site 8B, officially listed as State Hospital East.

⊳─౮౮౮─╢╟

Sandy Räker had been introduced to amateur radio in the seventh grade by her history teacher, who also managed the school's radio club. One day the teacher invited his class to the club's amateur radio station and made contact with a radio ham in China. Their conversation was entirely in English, which might have seemed international and impressive in itself to the German students. Numerous "QSL cards," printed confirmations that were exchanged after contacts as a courtesy, were posted on the wall.

While some of these were routine data-filled exchanges, many pictured mountains, lakes, famous tourist sites, and other attractions of the country. It was an unusual (and exciting) way to introduce his students to other cultures—by letting them actually talk with real people on the radio.

That demonstration fascinated Räker, but she was too young to get an amateur radio license. At the time, 15 was the minimum age to operate a club station, and being 16 or older was mandatory to have a station at home. Räker studied books and manuals, and as soon as possible she took the test and passed. Right away she became active in the school's club station. When she reached her sixteenth birthday, she got her own call sign, assigned to her family's address. In addition, she was awarded with a new bedroom. At first the only furniture included her bed, a study desk, and a bare spot for a future radio set. But her history teacher came over one day with a combination transmitter-receiver radio, one no longer being used at the club. Of course, an antenna was needed. So, together they constructed equal sections of wire, connected by an electrical feed line. Räker carefully did the measurements herself. Her teacher cut the wire and soldered the pieces to the connections. They brought the feed line through a window into her room, connected it to the transceiver, and soon she was on the air.

At the time, Morse code was mandatory for an amateur radio license in Germany, as it was in all countries, and Räker did fine as a new coder. Yet she found the voice option to be even more fun, since she was a relatively rare female operator on the different ranges of frequencies designated for amateurs. As a YL operator, short for "young lady," she was popular in a sea of male voices.

Räker's history teacher was an avid amateur enthusiast who enjoyed going after awards—usually for contacting all of the states or districts in different countries—but he didn't enjoy the pressure and organized chaos of radio contests, when the frequencies were crammed full of strong stations making quick contacts.

Sandy Räker needed an advisor for radiosporting, which appeared exciting to her. Yet contesting can be a daunting challenge to someone with no experience, so she took it upon herself to reach out. In the summer of 1991 she read about a "hamfest," where amateur radio enthusiasts congregated to look over new and used equipment, purchase odds and ends such as electronic components, listen to instructional seminars, and hear reports from groups who had ventured off to operate on remote islands. A regional hamfest was scheduled in Germany near the border with the Netherlands. She was determined to go by herself.

After talking her mother into it—no small feat—she bought a train ticket. "I just showed up," she said. Later she walked up to the registration desk, unsure of the exact lingo. "I would like to take part," she said. She ended up having a wonderful time. It was amazing; although she was a young girl in a predominately male setting, she fit right in.

One of the connections she made was with a group of fellow amateurs. They were older men from Helgoland, a tiny, windy resort island in the North Sea, off Germany's northeast coast. Manfred, one man in that group and a serious radiosporting enthusiast, became her mentor and advisor, a term in radio-speak called an "Elmer." This is a positive label, and almost all amateur radio operators have had at least one such figure in their lives. The Elmer guides the amateur into the hobby with a mix of technology and its own somewhat arcane vocabulary. Manfred invited her to a weekend operating event called a Field Day. This activity is held in nearly every country and is a combination social gathering, outdoor campout, emergency preparation practice, and spirited rapid-fire operating event in which individuals and teams exchange information reports.

Following the Field Day, with the help of her Helgoland mentor, Räker took part in her first big-time radiosporting event in the fall of 1991. It was the most popular of all individual contests, the CQ Worldwide. Since this was in the days before computers, while operating in the voice contest, she had to record each of her contacts by hand—and in pencil. These included the call signs of the station as well as the signal reports and the defined geographical zones worldwide. In a contest, it's important to keep track of each person you've contacted, since hundreds or even thousands of these brief connections pile up in a day. In addition to the sequential list of contacts, a handy quick-check called a dupe-sheet is used for this purpose. (You can contact another station only once on each frequency band.) Räker's mother sat beside her for the entire event and kept her daughter's dupe-sheets continually updated by hand.

Räker quickly discovered that being female provided an unexpected advantage. Her voice had a higher timbre that gave her an edge when she called another station in competition with others. At times the other operator came back directly to her call sign. But when it was impossible to get any call in the cacophony, he (usually a he) would ask, "The YL [short for young lady], please call again." That CQ Worldwide contest was an exhilarating experience for her.

Räker's brother also became interested in amateur radio and earned his license, but he gravitated to the technical side, building and soldering circuits and equipment. The technical challenge of designing the perfect electronic circuit with transistors and other components captivated him.

To her, however, there was magic in the way the shortwave signals *sounded*. She listened in fascination to broadcasts—some strong and booming like the "Voice of America," the BBC, Radio Moscow, and German stations. And yet she loved the excitement of getting on the air from her own station—making contacts, getting to know people all over Europe and beyond. Räker became reasonably proficient in Morse, but preferred operating with her microphone on voice (the word *phone* is used synonymously with *voice*), enabling her to practice the English language that she had been studying at school since the fifth grade.

The amateur radio signals from Western Europe and America were easy to understand, with stable transmissions, good voice modulation, and high-quality Morse code. These contrasted with the amateurs from the DDR, the Deutsch Demokratische Republik, as well as other Soviet Republics and economic bloc allies. They often hand-built their radios from scrounged components, and their signals often warbled and drifted around, with Morse code "chirps" like small birds calling for food.

Räker completed her undergraduate studies and then attended a police academy. Her academy education led to a job with the German equivalent of the FBI for twelve years in Frankfurt, West Germany's modern financial, banking, insurance, and business center. Later, to meet family obligations, she returned to her hometown in Lüchtringen, sixty kilometers south of Hannover, where she still works for the government in law enforcement as a criminal investigator.

Räker's teammate, Irina Stieber, was a tall woman who grew up on the other side of the then-divided Germany. Stieber worked in marketing and sales for *Die Sächsische Zeitung*, a German newspaper in Saxony. As a young schoolgirl, Stieber had heard about a class in Morse code. But her mother wouldn't allow her to enroll, saying she was too busy in her language and math classes as it was. By the fifth grade, Stieber had gone ahead and learned the code anyway. She found that it somehow satisfied her sense of mathematical order. Three years later, at 14, she earned her license.

---

The initial letters and numerals that make up a call, known as the prefix, are in themselves both interesting and the product of extensive international agreements. The original system was set in place in the early 1900s, with the major governments at the time getting access to what one might consider "natural" prefix letters. In the United States, the letter *W* previously had been used for stations east of the

Mississippi River, and *K* for those to the west. Certainly there were exceptions, such as KDKA in Pittsburgh, the first AM broadcast station, as part of Westinghouse Corporation, and KYW in Philadelphia. To the west, exceptions included WBAP in Fort Worth and WOAI in San Antonio, Texas. However, the vast majority of the *W*s were eastern indicators and the *K*s western indicators. So, the United States took these letters, as well as the *N* for nautical applications.

Internationally, over several decades and iterations of conferences, France took the letter *F*, Italy the *I*, and Germany the *D* (for Deutschland), while Russia got both the *R* and eventually the *U* (for the USSR), and Japan reserved the *J* sequence. Clearly there were insufficient combinations of main letters to cover every country. Great Britain, at the zenith of its colonial empire, took the obvious *G*, but also reserved *Z* and *V* for various colonies: Examples included *VU* for India; *ZS* for South Africa; *VP* along with different numerals for islands such as Bermuda and Turks and Caicos; and *ZD* along with different numerals for South Atlantic islands such as Ascension and St. Helena.

Germany was stripped of its colonial possessions after World War I and cleaved into two competing economic areas after World War II. Most of the *D* prefix remained with the part of Germany controlled by the Western powers. A new *Y* prefix was used by the Soviet-dominated DDR; therefore, Stieber's Y89RL call sign would have indicated to radio operators worldwide her East German location.

In the United States, Congress passed the Radio Act of 1912 as a result of the emerging interest of active radio amateurs. This act limited them to frequencies above the AM radio band at the time, or greater than 1.5 MHz. (It was thought that those frequencies were useless.) Subsequent experimentation by early radio amateurs led to the discovery in 1923 of HF (high-frequency) propagation via the ionosphere.

As a related matter of somewhat arcane interest, naval ships and aircraft use a similar international system of radio prefixes. An American-based airline uses the letter *N* to begin its radio-identifier call sign, since the letter *N* was assigned to the United States according to the 1919 agreement. An example is N8825 for a US-registered aircraft. Someone knowledgeable about the system (such as a compulsively obsessed amateur radio operator) can look at the call signs on airplanes on any airport tarmac and know their country of origin without delay.

On an additional level of factoid fever, the movie *Out of Africa* featured a wonderful scene with Robert Redford's character flying Meryl Streep's character in an old biplane. The plane's pre-radio identifier was G-AMMI, seen in huge letters on the top wing. At the time, that area would have been a British colony. Today, a plane registered in the same country would have a call sign beginning with 5Z, the current prefix allocated to Kenya.

---

Stieber's East German Elmer (mentor), who lived in Dresden, introduced her to high-speed telegraphy, in which the teenager achieved an expert level of "characters per minute," random transmissions of letters and numbers. Yet she was more interested in applying her advanced code-receiving ability to radiosporting at Y35L, the Dresden radiosport club station, where she became a whiz at Morse code amateur radio contesting.

 ⌁⌁⌁

After a short drive, Duffy's SUV reached the team's operating location on the grounds of the Medfield State Hospital. The former mental hospital had a bit of history. Several horror-movie scenes had been filmed at the location. A cult following is said to exist among those who believe in the supernatural. It's likely that Räker and Stieber hadn't been told of the numerous YouTube videos that claim to document eerie phenomena of images at night and strange sounds coming from the buildings. Yet at midday on Friday the place showed no signs of noises or apparitions.

The four passengers were met by the site manager, a brave (if you're into the paranormal) volunteer who had been there since the peripatetic "beam team" had erected the tower and large metal antenna. At first glance, the thing looked like an old-fashioned television aerial that had somehow mutated into gigantism—nearly fifty feet across. It perched atop the tower and was turned by a rotator motor, controlled inside the tent along with a direction indicator. This standard configuration was exactly the same for all fifty-nine sites and was complemented by two wire dipole antennas, each suspended near the top of the tower. The huge metal antenna, which covered three different frequencies, and the two wire antennas, each covering one frequency, made it possible for every team to operate on five different amateur radio "bands," thus requiring different skills and corresponding, roughly speaking, to daylight and overnight conditions.

The group carried their assorted luggage and cartons to the tent and stacked them outside. One by one, each container was unpacked and the contents carried inside and set up on the standard-issue table. Two complete amateur radio stations began to take shape, connected by a complex system that included a power monitor to ensure that no team exceeded the 100-watt limit. Computers and large monitor screens, along with Morse code keyers, headphones with swiveling boom microphones, and various filters and custom connections, completed the setup. It looked as if aliens had descended from space and constructed a mind-boggling communications node out of nowhere. And it was real!

A key piece of each WRTC station was an audio connection for the referee, who would sit in the tent with the team for the entire competition. The referee's duty was to monitor the operations to make sure that all transmissions were in English and that there would be no clues to the actual identity of the competitors.

To achieve the goal of anonymity, each team would be assigned a unique call sign that would be known only to the two team members and their referee for the duration of the contest.

A custom electronic wireless connection to the Internet had been issued to every team so that each site could transmit its ongoing score to the real-time scoreboard. Unfortunately, as it turned out, because many of the sites were in remote locations to optimize quiet radio conditions, some lacked Internet capability. Cell phone coverage often was the exception rather than the rule, and many reporting systems either never connected or simply couldn't stay connected that weekend. Therefore, updates for many locations had to be phoned in by the referee every thirty minutes or so.

Tim Duffy had been a valuable WRTC technical advisor in advance, and he had recommended the station configuration. He now took the lead as this station took shape. The women were a bit nervous. But anxiety changed to relief as the radios, Morse keyers, and complex computer networks all powered up without any glitches. Sandy Räker sat at her radio and made her first transmission using the microphone. To her excitement, her headphones nearly exploded with stations calling her. Irina Stieber did the same, using her preferred Morse code, and got identical results. The site appeared perfectly quiet in terms of electrical noise, which was extremely important. The two were ready to go!

Irina Stieber and Sandy Räker lead the German competitor delegation. Photograph: Michael Hoeding (DL6MHW).

# Chapter 3

# HAM WIDOWS' BALL

A major international event such as the WRTC brings together an interesting assortment of people: officials; volunteers; competitors; judges; a video team filming a documentary; radio, television, and print reporters with cameras, microphones, and notepads; visitors and spectators; spouses, friends, and partners; even this writer doing research for his book. At all previous WRTCs, tours had been planned before and after the radio competition itself. Of course, the 2014 World Radiosport Team Championship was no exception. Excursions to Boston began on Thursday before the opening ceremony so that everyone—competitors, referees, and guests—could take part. This schedule would continue on Monday when, once again, everyone could participate.

A sharp-eyed observer might have noticed something in previous championships: the lack of any planned event *solely* for the spouses or "significant others" of the almost entirely male competitors, referees, and other officials. Something for the nonradio people. One such observer did. His name was Bill Vinci. Vinci, an example of the extraordinary creativity and diversity of the San Francisco Bay Area, was the longtime and now married partner of Rusty Epps. Epps, as mentioned earlier, had been involved at the highest levels of radiosporting and competed in champion-ships since the beginning. He and Vinci were a familiar sight at these events, and Epps's rumpled, friendly manner was effective in resolving international radio issues as well.

Vinci was the opposite of his partner in many ways. In his mid-50s, he was a tall, high-strung person with demonstrative features, who exuded unabashed chutzpah. He would seem as much at home doing edgy stand-up in a comedy club as he was in this crowd of buttoned-down radio people. And this facet of his personality was evident in the Ham Widows' Ball, a wild and crazy special event he organized at the DoubleTree headquarters on Wednesday evening.

▷-◍-╫▶

In order to get more information on the Ball, I scheduled a sit-down with Vinci before the actual competition started. "I just got tired of coming to these things and seeing nothing, you know, *nada*, for the spouses and girlfriends," he said as he ordered a vodka and tonic at the DoubleTree bar. Vinci wasn't shy about say-ing exactly what he was thinking. "Get me a real vodka," he said to the grizzled bartender. "Grey Goose or Absolut. The best you got. I need something good."

The man behind the bar reached backward, and turned back to Bill. "Special selection, this one. Okay?"

"You got it," Vinci shot back. With a suitable drink now in hand, he leaned back precariously on the barstool and began to talk. "That's the way to do it," he said. "Life is too short for bad vodka. Rusty and I have been together for nearly thirty years. I've been to all these things. I'm not a radio guy, but I love to travel and meet people. Frankly"—he looked around the bar rather broadly—"I just got tired of all these guys getting together with all their com-plicated radio gear, all of them focused on the competition and all that stuff." He took a long, slow sip. "Ah! Good . . . thanks. This is on your tab, after all!"

He paused a second, tapped me on the arm, and went on. "But the wives and girlfriends, all those *wonderful peeps*—at some point all of us just sort of sat around with not much to do. It was particularly evident to me in Finland and Russia. So, I said to Rusty, 'Let's do something about it. Let's organize a special event for all these people who are left alone as ham radio widows during contests. We just sit there for hours while you guys are off wearing your earphones and staring into a radio set!' You know," he said, swiveling on his barstool to face me, "hello contest, *yada yada yada.*" With that he exploded in a raucous laugh. People in the bar looked our way. "Rusty told me to go ahead and come up with an idea. Said he would back me. So I called Doug Grant and suggested something special, just for the spouses and so on. Suggested I'd call it the 'Ham Widows' Ball.'" The name stuck.

Bill's delivery was razor sharp: He had perfect timing and hit just the right high notes and soft notes. "Grant bought it. Said to go ahead, then gave me a budget of fifteen hundred dollars. How about that!"

There were also donations from the Yasme Foundation and several individual radio amateurs. Vinci also reached out to a handful of wives of the organizers and competitors. Now a solid team, they had both reasonable financial backing and an official spot on the agenda.

Vinci knew his way around food. In addition, his art training at UC Berkeley had given him a strong sense of what he wanted in "the customer experience," as he put it. It was a "quality" over "quantity" plan. For the Ball, he specified a small number of up-market white wines and "really nice hors d'oeuvres, not any *cheeeesy* stuff." The way he pronounced "cheesy" made it clear that "the girls," as he referred to the guests, would have great munchies and wine.

The target audience of Bill Vinci's grand experiment didn't know exactly what to expect. "What in the world is a Ham Widows' Ball? And will there be music and dancing?" was overheard in quizzical conversations. But they did come, and in good numbers, walking past clusters of large black balloons in the hallway near the door to the conference room. After all, they *were* ham widows, weren't they? A final notice: A sign at the entrance door made it clear who was prioritized. A martini goblet was featured; however, the olive had been replaced with the international "not-bar" slashed across the image of a radio.

Inside the room, drinks were set up on a table by the door. At the back, serving tables were filled with upscale finger food. Two large posters were displayed on each of the two side walls: Art and Architecture and Food and Wine on one, History and Museums and Spas and Relaxation on the other.

"Good afternoon," said Vinci, microphone in hand, fully in control. "Welcome to the Ham Widows' Ball. Now you may not be sure what our agenda is, but *work with me*. You've noticed the posters. Please move over to the area that you most identify with. Go ahead—step on it!" He oozed positive exuberance. "*Move it, girls. Let's go!*"

Some of the women gathered at each station. Vinci's tone now was attractively flamboyant. "Okay, now introduce yourselves," he continued. "Tell where you live, who your husband is, and what his role is in the competition. Competitor? Referee? You got it. Take a few minutes to get acquainted."

Everyone began to speak. The ice was broken.

"Now, girls, over to your next preference. You know the drill. After that, go and find a seat." Soon the room was buzzing. Finally, the women moved on to two options of seating: regular tables and chairs, and hi-top stand-up tables.

Vinci took over the microphone again. "Now that we've met one another, or at least most of one another," he said, laughing, "I've got a question for you. Oh, yes, some nice bubbly goes to the winners."

Vinci then asked a series of questions. "How many of you have been to the Dayton hamfest with your husband?" Several hands shot up. "Now, how many have been to at least two previous WRTCs?" Fewer hands this time, but several. Some of these women traveled with their man no matter where he was going, at least for these big-time events. "Okay, come on up here with me," Vinci said.

The second phase of this series of questions involved re-creating a scene where the husband was completely occupied with a radio contest and how he paid attention, or didn't pay attention, to the spouse when she brought him food or had a question. One woman play-acted bringing her husband a tray with lunch. "Are you hungry? Here's something," she said. And then she mimicked his reaction. "He waved his hands and spoke in gobbledygook. 'Not now. Not now. I'm running Europe.'" Her arms waved around her as she imitated his rapt attention to the radio.

But that wasn't all. She went on to say that her husband didn't get up to go to the bathroom if he was having a good run (string of callers) going into Europe. "He keeps a half-gallon jug under his operating table."

"Oh my God!" shrieked Vinci in mock exasperation. It brought down the house. "That's absolutely dreadful. A *pee jar*." He handed her a bottle of wine and said, "You're a winner, dear."

After the group was seated, Vinci's face contorted into a dramatic pose. "Now, the final question. This is it. Who went out and used your husband's credit card because of a contest? I mean for spite. I'm not going to vote or I'd probably win," he said as the room cracked up.

A striking woman, Indonesian by birth and now married to an American, immediately stood up. "I charged five thousand dollars for an airplane ticket to go back and visit my family. To Bali! I met my sister there. I wasn't going by myself!" It was crazy now.

"Revenge! *Jihad!*" Vinci shouted. He had them in the palm of his hand. "That's it. We have a winner! You know, girls, we're all *so* alike. We're ham radio widows. Our husbands have a mistress. Don't make him choose!"

He then reached for a gaudy tiara. "If Mama's not happy, then ain't nobody happy," he proclaimed as he put the crazy crown on the woman's head, draped a glittering sash over her shoulder, and rested a bottle of wine in her hand. "Let's hear it for the Queen of the Ham Widows' Ball."

It was a blast. A smashing success. Everyone left wearing a smile and a string of faux black beads to signify their combined "widowship."

Bill Vinci at the Ham Widows' Ball. Photograph: Charles (Rusty) Epps (W6OAT).

*Chapter 4*

# QUALIFYING—IT'S HARD!

Qualifying to compete in the WRTC is tough. People go to extreme degrees to win a spot for their region of the world. Most don't have a competitive home station, one that would allow them to remain and battle it out, since this type of setup usually requires multiple towers with plenty of antenna "horsepower."

John Barcroft was a retired broadcast engineer from southern California. He had picked out a special call sign, K6AM, to reflect his early days with AM radio stations in and around San Diego. Barcroft's background, like that of over half the competitors, was technical. He started at small 250-watt local stations, working as an on-air all-night Top 40 rock 'n' roll disc jockey. Following college, armed with a degree in electrical engineering and practical in-studio experience, he capped his career as chief engineer for fourteen broadcast stations on both sides of the border near San Diego.

Barcroft developed and followed a clear but complicated roadmap that earned him one of the two qualifying positions assigned to the southwestern United States. He took part in only two contests from his own station, both using voice and both with the 100-watt category, where he believed that his chances to win or place near the top were more favorable. In three other events he

used a fabulous setup in rural San Diego County, where he operated alone in the high-power category. Although he didn't win any
of these, he placed very well. Barcroft's qualifying credits were bur-
nished nicely by nine additional team efforts at this superstation,
and two of these resulted in number-1 finishes with maximum
point credits toward the WRTC. In addition, six times he flew to
the Cayman Islands, where he operated three contests alone and
three others as part of a team. He continued to rack up the air miles
traveling to the Dutch island of Curaçao, once operating alone and
twice with a group.

By now you're probably both impressed with Barcroft's tenac-
ity and glazed over with the effort required. Bottom line: Qualifying for the WRTC is neither easy nor simple, and certainly not
cheap if travel and time are factored in. His twenty-three different
commitments over three years made that clear. Yet Barcroft had
identified a path through a labyrinthine process. If that weren't
enough of a challenge, he also had to work cooperatively with
one of the undisputed younger stars of radiosporting, Dan Craig,
N6MJ. Together they divvied up the available seats, or operating opportunities, at the big San Diego station. They developed a
schedule to avoid scheduling conflicts, sometimes teaming up in
joint multi-operator categories, other times going solo. For this
plan to work, each one had to enter and win nearly every specific
qualifying contest in the rugged high-power category.

Barcroft's search for a WRTC operating partner ended when he
found David Hodge, N6AN, a lanky television broadcast engineer
who had spent much of his career outside the technical field as a
professional musician. Hodge plays the French horn. (He'll correct you with the clarification that "a horn is a horn, I prefer horn
player.") At any rate, he was first-chair (French) horn in a major
classical orchestra in metropolitan Mexico City for years.

▷꧀꧀꧀─╟╟

This complex and dedicated commitment to qualify for the WRTC wasn't restricted to John Barcroft and Sandy Räker. In Japan, "Don" Kondou became an absolutely driven man. Like many radio hams, he had a technical degree—electronic physics, in his case—and had worked originally with the generation of industrial power, but now was employed in telecommunications.

Kondou had been away from amateur radio for twenty years. His first wife hadn't supported that hobby at all. Because his career and new second marriage kept him busy, he had little time for the shortwave passion he'd followed as a teenager. He and his new wife took up hiking, mountain climbing, and outdoor activities. Yet one day she discovered an old storage container with lots of outdated radio gear. Finding the equipment interesting, she encouraged him to get back into the hobby.

"Really? Can I do that?" he asked.

So much time had passed that his old equipment was out of date. But with his wife's active support, he began to practice his nearly forgotten Morse code skills during lunch breaks at work, using a simulator on his computer. He attended the Ham Fair in Tokyo, a major summertime gathering sponsored by the Japan Amateur Radio League. The fair was complete with vendors, new equipment, and a flea market for used radios and all sorts of paraphernalia. The flame was rekindled.

At the Ham Fair he bought a fiberglass rod and brought it back to his apartment, which was located on the street-level floor of a high-rise building—not exactly the perfect location. He configured it for a special type of antenna called a quad, which produces an effective antenna pattern but looks like a strange, almost kite-like periphery of wire around an X-support. He managed to get it constructed and extend it away from his "low-rise" window, if only by a few meters. Unbelievably, the spindly antenna worked. In fact, it worked so well that he made another one. Now there were two of them: the original facing in one direction, and the new antenna facing 90 degrees away, since his apartment sat at the corner of the building. All of this probably raised some eyebrows, at least from the other apartment owners. But it got Kondou back on the air in the spring of 2011, and he made several contacts in a major American contest on voice. When he sensed the return of a long-suppressed competitive twinge, he reached out to a prominent local radio enthusiast. This opened the door to guest operation at the friend's station near Tokyo.

Kondou and his friend cooperated in several domestic Japanese contests. The spirit was back. At the end of May 2011, they teamed up and made a serious entry into a popular worldwide Morse contest, one in which everyone could contact everyone else, with the "diversity factor" being the call sign prefixes. The twosome did well, and Kondou's long-dormant Morse skills came back nicely— it was like riding a bicycle after a long break. By now, the thrill of radiosporting was exhilarating. Was the WRTC a possibility? So far, he had a few modest finishes but overall was near the bottom of his Asian qualifying zone.

It was clear that his own strange crow's-nest setup at home couldn't be competitive in the contest world. Using his friend's knowledge of the radio community, Kondou met another ham who owned a large station—somewhat of a rarity in Japan. The man's station would have been a first-class setup anywhere, but it was especially so in that densely populated country. It was located in a hilly area far from any city, and the owner used it by himself mainly to "chase DX." (DX is a term for distance, referring in particular to a faraway, unusual, hard-to-reach, or politically isolated location.)

They worked out an agreement that allowed Kondou to use the big station for Morse code competitions, since the owner preferred voice. But he needed other alternatives, because a third of the qualifying window already had passed. He found the answer on the Internet; someone in the United States was looking for operators to join a team at an American-owned station on the Caribbean island of Bonaire.

For many years, the island had been the base for Radio Netherlands's powerful relay station to the Americas, and was known as an outstanding location for scuba diving as well as shortwave radio. In addition, Bonaire is located just inside the northern geographic edge of South America. So, contacts with the ham radio population hotbeds of the United States and Europe would be more valuable, since they cross continental boundaries. Kondou jumped at the chance and reserved a spot in a "big daddy" of the radiosporting world, the CQ Worldwide voice contest, at the end of October. The team finished second to a group in Brazil, but Kondou racked up a tidy point total toward qualification. He'd also broken into a circle of people who liked to compete from the Caribbean. In 2013 he traveled to Antigua, again in October, and then wrapped up his effort to qualify from the Japanese hilltop superstation, since the owner had a business conflict and the setup was available.

His reemergence in a hobby that was supported by his new wife, his efforts in Japan, and his international travel and teamwork resulted in a strong finish—but only number 2. Fortunately for him, the top individual contester in Japan wasn't active outside the country and showed no interest in claiming his number-1 qualifying score. In fact, he didn't apply for the WRTC 2014. As a result, Kondou-san, who had been completely out of the hobby and unknown in the radiosporting world three years earlier, became the qualifier!

Now he needed a partner. He asked his radio friend Tack (radio hams use first names), who seemed to know everyone involved in amateur radio in Japan, "Who's the best guy?" Since the modern WRTC format is based with teams in tents, running on generators in somewhat isolated locations, it reminded them of Field Day operations. This led Tack to recommend Hazuki-san, a mechanical designer of factory automation robots and a five-time champion in somewhat unique Japanese domestic contests. The 29-year-old Hazuki was very good on Morse code and said he was "okay" on voice. But since the WRTC would be a Morse-dominated competition, Kondou-san extended the prestigious invitation. Hazuki accepted immediately. The two found a synergistic love of radio-sporting in addition to mountaineering. (Hazuki had once lugged heavy radio gear all the way to the top of Mount Fuji in search of an optimum location for one of the Japanese competitions!)

They began to work on the admittedly daunting configuration now de rigueur for the WRTC. Neither one was a broadcast engineer, and they didn't have a figure like Tim Duffy to guide them as a senior mentor. They asked around, got advice, tried things, and fixed problems that inevitably arose. They practiced first with the complete configuration at Tack's station in September 2013. Later they made a serious dry run at the big rural station in March 2014 in the Russian DX contest, a twenty-four-hour wide-open affair that permits both voice and Morse code. All of the filters and the special "Tri-plexer," which permits both transmitters to use the same antenna simultaneously, worked perfectly. There was no interference at all between their two stations.

By then, only one final uncertainty existed. As a dedicated Apple computer user, Kondou had attempted to develop his own software program to run the contest. He didn't want to use a PC but quickly decided his efforts wouldn't meet the rigid standards. Someone mentioned that a Mac-based program had been developed by a likewise-dedicated Apple aficionado in New Hampshire. This led him to a new program called SkookumLogger. Kondou obtained a copy, found that it worked for him, and obtained late approval from the organizers to use it. At last he and his teammate were ready for action.

L to R: John Barcroft (K6AM), David Hodge (N6AN), and referee Andrey Stefanov (LZ2HM). Photograph: Bob Wilson (N6TV).

L to R: "Don" Kondou (JH5GHM), and Hajime Hazuki (JA1OJE), with referee Kristjan Kodermac (S50XX). Photograph: Bob Wilson (N6TV).

# Chapter 5

# FINDING SITE 15W

Doug Grant had steered this WRTC all the way, starting shortly after the closing ceremony in Moscow when the Americans proposed—and then won—the right to host the championship in 2014 over a competing Bulgarian offer. His decisive, never-in-doubt style and a topflight leadership organization of experienced hams, with mostly northeastern US roots, could match nearly any corporate organization in dedication and efficiency. On Thursday night, to a packed house, Grant declared the WRTC 2014 officially open. Everyone was there: competitors, officials, spouses, and guests. Every seat at every table was reserved, with the additional crowd jammed around the periphery of the hotel's largest conference room. The mood was controlled excitement.

At the station draw the next morning, the atmosphere was different. Formality and ceremony transitioned to efficiency and business. Site selection mattered. At the WRTC in Moscow four years earlier, the sites had been nearly identical, laid out in a grid across three clusters within a flat open space measuring thirty by forty kilometers. The competitor locations were as equal as they were ever going to be, spaced 500 meters apart in geometric patterns.

However, that exact system wasn't possible in New England. As Grant noted, the land had been settled centuries ago. Flat spots were tilled for crops, and homes were built where most practical.

A three-year program to identify more than sixty sites—a project of maddening complexity—resulted in an arc of locations sweeping from the New Hampshire border northwest of Boston, over to the west and southwest of the city nearly to Rhode Island, and finally to the southeast, almost at the beginning of the bridges to Cape Cod. The names reflected both Native American and colonial history: Keyes Nissitissit Meadows, Devens Lake George, Wompatatuck Central South East, Massasoit Cranberry Bog, and the lengthy Myles Standish North Three Cornered Pond East! At the smallest clusters, only one or two WRTC sites were approved, while the Devens cluster encompassed ten. At the largest of all, an hour and a half to the south of Boston, seventeen WRTC teams were tucked here and there in the Myles Standish State Forest, a vast tract of scrub pine, small lakes, and occasional heavy stands of hardwood.

Manfred Wolf and Stefan von Baltz had drawn site 15W, and immediately after the official photographs by "TV Bob" Wilson (a.k.a. N6TV), the two contestants, along with referee Wes Kosinski, picked up the information packet with directions to their site.

Suddenly, in that ballroom at the DoubleTree, my name was called. Actually, it was my call sign, which was a more familiar identifier to most in this group. "N3BB, come to the drivers' area, please."

I had volunteered to be a team driver, and now I was being assigned to the German team. I knew they were good—very good, actually—and it seemed a good way to "get in the game" and embed myself in the action. Their location appeared to be somewhere in an isolated rear section of the Myles Standish State Forest. A notation indicated that the map itself held "insufficient detail to locate the site" and referenced us to additional pages, indecipherable at first glance. This extra section indicated what appeared to be small forest roads that spread out in a spiderweb fashion in indistinct light gray lines. GPS coordinates to the third decimal place were included.

Thirty minutes later, Wolf and von Baltz emerged from the elevator into a sea of people. Their small mountain of gear was mounted on a full-size hotel trolley. Excitement and a palpable tension had replaced the protocol of the lengthy selection procedure, which in turn had superseded the formality and, to some degree, pomp of the opening ceremony.

At Boston's Logan Airport, the choices had been limited at the rental area, and as a result my car, and therefore the team's transportation, was a somewhat racy-looking Dodge Charger. The trunk area was huge—a pleasant surprise—and the rigid, heavy containers, designed to protect fragile radios, fitted nicely into the far reaches. Other, more flexible items were wedged in and around until only three boxes remained on the trolley. Von Baltz worked one into the last possible trunk cranny, and the final two made the trip wedged between the two Germans in the rear seat. Kosinski rode up front in the passenger seat.

Finally, I pulled out of the DoubleTree onto a tricky side-access road. It seemed like we had to go one way in order to end up going another, an assault on the sense of orderly direction. The access road crossed over Route 9 on an overpass. Now we headed the opposite way, looking back at the hotel across the busy highway, and within a quarter mile merged onto I-495 heading south. For nearly sixty miles it was smooth going. A few miles after passing the I-195 exit to Providence, we exited onto Massachusetts Route 58. Then began the fun of traversing New England back roads.

Somehow, the Polish referee, making his first trip to America, became the navigator as two excited Germans looked over his shoulder. We passed several large cranberry bogs and then a massive processing plant for the fruit.

"We're supposed to follow Route 58 until we get to Tremont

CONTACT SPORT

Street," I said, glancing over at the directions, which Kosinski was holding for me to read.

"I think so," came the response in three European accents.

"After that, the directions say to turn right onto Cranberry Road," I said, trying to act like a driver who would get them to their spot. There it was, Tremont Street, just ahead. So far, so good.

These old towns had classic street signs, reminiscent of the 1950s. The lettering often was faded—you had to pay attention. It reminded me of the comment I'd heard in rural Pennsylvania: "If you don't know where you are, you shouldn't be there in the first place." Cranberry Road certainly was aptly named, and we soon left a quaint residential neighborhood and meandered between what appeared to be large, flat, sunken fields, two to three feet deep and crisscrossed by small drainage ditches. They were working bogs, which became clear when we passed a large area with commercial sprinklers. After a mile or so of more bogs, now on both sides of the road, we gained elevation and passed through a grove of trees.

Cranberry Road ended at the state forest headquarters, an appropriately woodsy-appearing structure. At this point the written directions referenced backup sheets depicting magnified maps (and requiring, perhaps, a magic decoder ring) of forest roads. The font was small and difficult to read, and every other name seemed to include the word *pond*.

A car parked just ahead had a telltale mound of large radio containers and other luggage that prompted me to pull alongside. The driver, a former air force fighter pilot, was one of the two operators of the Hawaiian team. He was talking on a cell phone and motioned us to wait. Then, putting the phone down, he said, "Just got it . . . had to call for directions. We're a bit ahead at the first spot. Good luck. This place looks huge."

That much was obvious. We watched the car pull out and slowly drive up the hill into the trees. No one said much. Finally, I noticed two uniformed forest rangers sitting under a group of trees near the building. "Let's check with them," I said, and parked our car. Kosinski and I walked over to the men, who were surrounded by diagrams and maps that were spread out over a wooden table.

"You guys with that radio group?" one of them asked. It was as if we had stumbled upon an information oasis in the middle of nowhere.

"We are," I said. I handed them our map. "Here. Site 15W."

The ranger squinted and seemed to hesitate. "That one's out there," he said. "I can show you where it is." He whipped out a forest service map and proceeded in rapid-fire proper Bostonian to rattle off names. "This road"—he motioned over his shoulder—"is Fearing Pond Road, but the bridge over there"—he nodded to the right beyond some trees—"is out."

I tried hard to grasp the entirety of the situation.

"So, you'll want to go to the left and drive around the big pond. Just take Lower College Pond Road." Now he nodded to the left. "Go up that hill and keep to the right. Just keep turning right. You'll end up back on Fearing Pond Road." At this point, he swiveled his head and nodded back toward the right, to the bridge that was closed somewhere over there. "When you get back on Fearing Pond Road, keep on it and follow the signs to Camp Squanto. That's important. You want to go toward Camp Squanto. You can't miss it."

Kosinski seemed baffled by the high-speed conversation—not surprising. I wasn't much better. Back in the car, off we went up the hill and kept to the right on Lower College Pond Road, which continued for several miles. A large mountain lake—apparently a "pond" in New Englandese—showed through the trees to the right. Suddenly an intersection appeared ahead.

"Just keep to the right." I kept repeating the phrase in my head.

I turned right onto Halfway Pond Road. We were doing thirty miles per hour now and made another sharp right ("Just keep to the right") at the next crossroads onto Upper Pond Road. Clearly the word *pond* was a part of many of the forest roadway designations.

Another intersection appeared after a mile. The only possible route was to the left, and a "Road Closed—Bridge Out" sign indicated the way back to the bridge outage that required this detour. Finally back on Fearing Pond Road, we continued until the eponymous pond became visible on the left, encircled by a one-way lane named appropriately "Circle Drive."

The Germans, along with their Polish referee, now seemed a bit restless. They weren't alone. Then a true godsend appeared—a weather-beaten arrow to the right, pointing to an older, washed-out road: "Camp Squanto."

I managed a quick turn. The Charger, showing some impressive nimbleness, lurched to the right. Old asphalt pavement, apparently Cuttersfield Road, according to a miniscule sign that had been too long exposed to the elements, made a sharp right-hand bend and then straightened out onto a grassy expanse. Things looked better now. Half a mile farther, a sign appeared on the left: WRTC site 15B. Off the road, a tower and antenna loomed over a tent. Clearly this was one of the team sites. Maybe we would locate "our" site after all.

Then there it was! Almost immediately across the road, another sign had been driven into the ground: WRTC site 15W.

One more comment from the forest rangers now came to mind: In addition to "Keep to the right" and "Follow the signs to Camp Squanto," the ranger had inspected the site map for 15W carefully and said, "Oh, that one. You probably won't be able to drive a car back there."

Great, just great. I parked the car in the grass across from the site 15W sign and relayed the ominous remark to the team.

They decided to walk to the site, as the word *road* didn't seem to apply to the dirt swath a foot or two wide. Maybe it was a bicycle path, since there were no signs of horse droppings. After several hundred yards and in the middle of thick woods, the trail dropped sharply downward.

"This looks bad," said von Baltz. "We don't want to be in a hole."

Ahead loomed an area that clearly would be a wet-weather stream, and a muddy mess for days after a rain. Now it was caked bone-dry. On the other side, the ground began to rise and regained all the lost elevation over another 200 yards until the path emerged into a large open field. Someone had cut out the trees at some point. The area apparently was mowed periodically—baby pine trees two feet high with trunks an inch thick had started to regrow. We continued into the clearing and saw the WRTC antennas, along with the tent that was set in the far end along a tree line. Two men waved as we approached, undoubtedly relieved finally to meet other human beings—especially the team that would be competing from their site. Michael Bennett and his son, Derek, had been at the site for nearly two days, securing it following the beam team's installation of the tent, large tower, and antennas.

The father and son were pleased to see the team, and the feeling was mutual. It was clear that the trail was drivable, at least with the husky pickup truck the Bennetts had parked along the far tree line. With this information I volunteered to retrace the nearly third of a mile back to the road, to see if our sporty Dodge Charger with a full trunk could navigate the so-called entrance to the site. As I walked back—alone along the trail, or whatever it was—it was certain that someone or something had cleared the path. Side branches that normally would have obscured the trail had been lopped off.

A few of the larger limbs were substantial, an inch or more in diameter. These would rub against any vehicle passing through. That was certain. But if I could keep the car centered, the path might be wide enough . . . maybe with a few marks on the paint, maybe not. The decision was never in doubt. Back on the road to Camp Squanto, the car started up with a flare, and crossed the road into the narrow opening (otherwise known as a forest pathway). Certain portions were absolutely glen-like with lovely ferns. But in other sections the stiff ends of the branches rubbed against the side panel and made squeaks and high-pitched abrasive noises.

And small stubs of decapitated baby pines grated against the car's undercarriage panels, making a combination of funky thumps and throaty brushing sounds. At the slope down to and across the wet-weather swale, the road was rougher, with small stones and chipped bedrock. But nothing was too bad, even for the American muscle car with a large storage trunk and well over 200 pounds of complex radio and computer electronics. At last, with a bit of personal triumph, I arrived and parked the car with its trunk end facing the competitor tent. Elvis, in the form of the precious payload, now was (almost) in the building.

The father-and-son site team were a good example of the rich volunteer spirit of the entire event. As outdoorsmen they were well equipped and had set their auxiliary tent and campsite a hundred feet from the WRTC tent, where they could make sure no one bothered the standard-issue gear. Like nearly all the volunteers, they were licensed amateur radio operators and were "on the air," but on a modest scale compared with the prominent competitors and referees. Their tent was situated at the back tree line, and the sixty-foot pines and hardwoods presented soothing shade half the day. Their heavy-duty pickup truck was parked nearby, with its bed full of cooking stoves and other camping necessities.

After introductions, von Baltz, Wolf, and their referee checked out the site, with the Bennetts and me in tow. The clearing was two or three acres in size and, according to Michael Bennett, was used as a pheasant and grouse hunting area. The area for the tents and campsite had recently been mowed nearly to the ground, although a few remaining pine "trunklets" held on stubbornly. People needed to wear shoes and watch their step.

Bennett described the beam team's setup process. A huge X pattern had been close-cut in the field a hundred feet away from the competitors' tent. The tower and antenna were assembled separately in this clearing, then one end of the tower was raised and the antenna bolted into place above a rotating motor. Two smaller

wire antennas were affixed near the top, as well as four supporting guy wires. Following that, the entire contraption was raised,
sphinxlike, in one clever swoop using a specially developed coun-
terweight that had been deployed at each of the identical fifty-nine
setups. The volunteer beam team had worked out this technique,
and the entire procedure had taken only a bit over one hour.

"I watched the whole thing. They had it down pat," said the
elder Bennett.

Von Baltz thought it was important to confirm that the large
antenna was aligned properly—that is, that it actually pointed to
Europe just as the indicator screen inside the tent said it did. From
Boston that direction is around 45 degrees on a great circle map. He
walked slowly as he referred to a slick compass app on his "handy,"
as cell phones are called in Germany, pointing the phone this way
and that while Kosinski sat in the tent and moved the rotator control. "Ah, it's good," the tall von Baltz pronounced, then turned
his attention to the placement of the two ends of the wire dipole
antenna for 7 megahertz (MHz). That frequency band would be a
key workhorse into Europe as soon as the sun went down.

The German team confirmed with their referee that they were
allowed to reposition the ends of the wire antenna, which were
easily accessible. Then the three of them proceeded to move each
end thirty feet or so until the antenna was broadside, more or less
facing Europe. At this point the two endpoints of the wire were
located at the southeast and northwest so that the main signal pattern went to the northeast as well as to America's southwest. "Now
I can see Europe over there, right through the trees. It's better,"
stated von Baltz with confidence. The other dipole, for the 3.5-MHz
frequency band, wouldn't be as critical. But it was important to
offset it from the other wire antenna so it would have minimum
electromagnetic interaction with the primary wire antenna. The
team, with their referee and me working together in the spirit of
international cooperation, trooped around the edges of the mowed

section and moved the final two ends around to the northeast and southwest. With a final reference to his "handy," von Baltz did one more check and smiled broadly.

"*Alles ist in Ordnung,*" he declared. Everything is in order.

The station now took shape. Like the German women, the guys had lots and lots of equipment. Manfred Wolf toiled beneath the table to position huge filters that were designed to minimize signals from one radio interfering with the other. One of these was nearly three feet long and nearly a foot wide. Ten minutes later, he had an impressive grouping that looked like neatly cut chunks of metal, laid end to end, that covered the rear area of the canvas tent bottom. Within two hours an impossibly disparate collection of cables, radios, boxes for this and that, two computers, two huge LCD monitor screens, and miscellany sufficient to fill up the entire operating table sat in place.

The operating area itself looked similar to a military field installation, with the two giant color monitor screens dominating each of the operating positions. To someone standing in the tent, after having snaked through the small entrance flap, the picture was one of space-age paraphernalia, with bright high-definition computer screens and flashy Japanese radios, each with its own impressive readouts for the frequency, and a pan-adapter screen showing the strength of incoming signals. The dramatic high-tech picture was almost entirely visual, because all of the actual incoming radio audio was routed to the individual earphones, as well as to the circuits to Kosinski, who would somehow manage to listen to both guys, either individually or simultaneously, without losing his mind.

Von Baltz sat down at his operating position on the right, put on his headphones, and cautiously made a test transmission. He moved the transmitter to all the frequency bands the competitors would use and declared to Wolf, "*Es geht.*" It works.

The two men ran both stations through their paces. Everything checked out on both Morse code and voice operation. Everyone started to relax. The site and the setup looked good. Although they were in a forest, the proverbial coast was clear. Or was it?

Packed and ready to roll! L to R: Wes Kosinski, Manfred Wolf, and Stefan von Baltz at the headquarters hotel. Photograph: Jim George (N3BB).

Jim George (N3BB), getting directions to find the site with the Myles Standish park rangers. Photograph: Wieslaw (Wes) Kosinski (SP4Z).

L to R: Stefan von Baltz (DL1IAO), Manfred Wolf (DJ5MW), and Wes Kosinski (SP4Z) getting ready to walk the path to the isolated site 15W in the Myles Standish State Forest. Photograph: Jim George (N3BB).

L to R: Stefan von Baltz, Wes Kosinski, and Jim George on the original walk to site 15W in Myles Standish State Forest. Photograph: Wieslaw (Wes) Kosinski (SP4Z).

Very first view of the forest clearing of site 15W. Photograph: Jim George (N3BB).

Setting up at W1P. Lots of wires and meters. Photograph: Jim George (N3BB).

L to R: Manfred Wolf (DJ5MW), referee Wes Kosinski (SP4Z), and Stefan von Baltz (DL1IAO) at setup Friday. Photograph: Jim George (N3BB).

Tower and antennas at W1P site. The antenna is pointing at Europe over the tree line. Photograph: Jim George (N3BB).

*Chapter 6*

# GETTING SET

By this time other teams had arrived, and now test transmissions were flying over the air. Of particular concern to every team was the specific configuration of WRTC sites located in this far-reaching portion of the state forest. Site 15W was one of seven, set in the geometric grid pattern shown below.

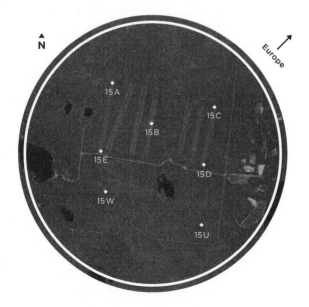

These sites seemed flat, which was good since no one wanted any upslope or rise toward Europe. Most of the big antennas were out in the open, considered better because of possible interaction with large trees. But that point was debatable. The German team's site was fairly open, but the one nearby tree line *was* in the direction toward Europe. Yet, with seven world-class radiosporting teams within roughly 1,000 meters, the team's major worry was strong signal interference, since the 100-watt signals could overload neighboring radios. Site 15W was positioned at the southwestern corner of this location; therefore, when the teams were beaming to the northeast, toward Europe, the main lobe of signals would be aimed away from them. That was a positive.

To the Germans' immediate east, although they didn't know it at the time, was the team most oddsmakers would have pegged as the favorite to win. According to the radio gods of fate, this team had drawn K1A as their call sign—smooth to say on voice, and rhythmic and snappy on Morse. Their site at 15U was over 500 meters away, completely invisible through the dense pine and hardwood trees. When both teams pointed their main antenna at Europe (which would be a large majority of the time), the signal patterns would be "off the side" to each other, since the strongest part of the transmissions would fly off overseas.

The men at K1A, Dan Craig and Chris Hurlbut, were at a prime age: plenty of youthful stamina, as well as experience at world-class radiosport competitions. Craig hailed from southern California and worked as a pharmacy auditor. He had cooperated with John Barcroft, K6AM, to line up specific guest seats at the superstation in the mountains of San Diego County and other locations. His resume was so strong that Barcroft and others were, in practice, shooting for the "other" qualifying position in the ham radio–rich zone of the southwestern United States. Craig had nearly won the three previous WRTCs, working with Hurlbut and Dave Mueller. (This year, Mueller was only 500 meters away on the other side of Cuttersfield Road as part of the network CEO's sponsored team.)

A great deal of radiosporting power had been placed in one small area. On a metaphysical level, perhaps mischievous radio deities were watching from an ionospheric Valhalla while listening to a Morse code version of Wagner.

Hurlbut, like Craig, was neither an engineer nor a professional in communications. The Alaska native now worked as a supermarket manager in the "lower forty-eight." He got his "radio fix," as he called it, in short bursts on weekends, guest-operating as a "hired gun" at other stations. The pay was paltry—some food and drink and a place to sleep—but the radioing was good. Hurlbut and Craig had given the Russians a run for their money in Moscow, falling only 4 percent behind the winners in a strong field.

⊳─ꞏꞏꞏꞏ─╫⊩

The third portion of this triumvirate of highly skilled radiosporters set up just across Cuttersfield Road at site 15E, where Tom Georgens and Dave Mueller were assembling their complex gear. Georgens had no amateur radio setup at home and often operated from the island of Barbados in big radiosporting events. He owned a modest house there and had built a competitive station that "got out" well to both North America and Europe. His track record was top ranked.

Georgens had the appearance of CEO casual: knowing smile, baseball cap, rimless glasses, self-confidence to spare. From his Barbados location he excelled at running (contacting) US stations on voice, amassing incredibly huge yet accurate numbers in never-ending pile-ups (jams) as the Americans tried to reach what often was the only station active from the Caribbean island. (The word *running* is used to denote contacting other stations rapidly one after another. A *pile-up* happens when large numbers of callers try to reach a single station at the same time.)

Mueller had cut his teeth in contesting in central New Jersey, where he was mentored by the Frankford Radio Club, a large radiosporting group centered in Philadelphia. He was now 42 and served as an officer in the US Coast Guard. His assignments had included the northeastern United States, Florida, and Guam. Now he and his family were being transferred to Honolulu as the next stop in his peripatetic career. Because of the move to Hawaii, Mueller originally had planned to sit this WRTC out and perhaps fly to Boston as an observer. That way, he still could spend time with the contesting crowd, many of whom he hadn't seen since four years earlier in Moscow. But when Georgens asked him if he could work on a sponsored team, giving it a go at the highest levels, Mueller couldn't say no. On top of that, since Georgens had his entire WRTC setup from Moscow, Mueller needed to bring only a radio, headphones, and a Morse key.

Mueller had been responsible for the station integration when he and Dan Craig competed in the Brazilian WRTC 2006. It was amazing that he made it to Brazil at all. His plane tickets had become worthless when Varig, the Brazilian airliner, went bankrupt just prior to the competition. Fortunately, friends and supporters had stepped in at the last minute and purchased tickets for him from another airline. So—in one of those cases of events falling into place just right—here he was again at a World Radiosport Team Championship just across the road from Craig and Hurlbut. All of them had been extremely close friends for fifteen years. In fact, it was Mueller who had recommended Hurlbut to Craig for the Moscow competition, since he and his family were being transferred to Guam at the time.

Their referee was Chris Burger, a South African and a veteran of several WRTCs. Burger had Boer ancestry, chiseled features, and a five-o'clock shadow at noon. His technical background in South African military circles provided a certain intrigue. Yet, at the same time, his wicked sense of humor and distinct accent kept everyone loose.

Mueller and Georgens completed the assembly of their complex equipment glitch-free. Their location at site 15E was just off Cuttersfield Road in a grassy field—no trees at all. In fact, you could walk from the road to the tent in one minute, as opposed to the difficult access into the grouse and pheasant hunting area where Wolf and von Baltz were located.

⊳〰〰⊩

In order to provide an "arena" for the world radiosporting teams, the specific WRTC competition was inserted into an ongoing major summer contest called the IARU (International Amateur Radio Union) HF Championship. The IARU was perfect in many ways: It was held in the summer (suitable for tents and outdoor setups), it included both voice and Morse operation, and it lasted twenty-four hours. Many thousands of regular radiosporting enthusiasts took part each year. The WRTC could be embedded into the IARU contest every four years nicely.

In both the IARU and the (embedded) WRTC contests, voice (or what is called "phone") and Morse code contacts counted the same in terms of points per contact. Although voice was a little faster than Morse code, this advantage could be minimized with the very high code speeds produced by these elite operators. In general, phone operation could produce 300 contacts or more per hour under ideal conditions, while 200-plus was terrific for Morse code. You might have concluded that the strategy in the WRTC would be voice-centric. But traditionally, Morse "gets through" better in marginal conditions, and historically, many contest operators have preferred to use Morse rather than voice.

In addition, a recent technical development had tilted the playing field even further in favor of Morse: the so-called Skimmer. It's been said that the Internet changes everything. To underscore that point in radiosporting, clever people had written software that detected and decoded Morse code. Selected receivers around the world had been set up with detection software and Internet connections. These immediately "reported" both the call sign and the frequency when an amateur station sought a contact by calling "CQ." This technique was perfect for Morse code contests, where operators called CQ quite often.

> The term *CQ* probably (no one is sure) began in the early Morse days before voice technology. There is no written record to confirm the history, but CQ gained recognition as signifying "Seek(ing) You." The shorthand took hold both on Morse and on voice, and a general call goes something like this: "CQ, CQ, this is [insert your call sign]."

As a result, when the WRTC Morse operators—or anyone else, for that matter—called "CQ," information was posted immediately on the Internet. Radio operators worldwide were alerted—from seasoned radiosporting veterans to newbies just experiencing their first opportunity to contact someone in another country. As a bonus, the high-flying speed demons' call letters appeared on a computer screen with their operating frequency. Bottom line: "The Internet changes everything" certainly applied in this situation and, somewhat counterintuitively, top teams would need to prioritize Morse code in order to achieve the most radio contacts. The Skimmer, along with the ability of Morse code to get through interference and noise if signals were marginal, made this strategy necessary.

In the hallway conversations at the hotel, the chatter in the bar, and the best guesses made in the lobbies, people were vocal that it should be possible to make over 4,000 contacts, maybe even 5,000, during the contest—within twenty-four hours! This prognostication was based on a key change in the rules: For the first time, the WRTC would allow both operators—both team members—to transmit at the same time. In the past, while one person "ran" (contacted) stations and the other searched for new multiplier stations, the "runner" would have to stop and go find the target multiplier, contact that operator, and then return to where (he hoped) he still had a group of stations waiting for him. Now, however, it would be wide open. Both operators were left free to make contacts as quickly as possible, but the team would have to make time for the multipliers as well.

To make matters better—or worse, depending on how you looked at this wide-open affair—the New England organizers had changed the rules for multipliers as well. Previously, the definition of *multipliers* had included a modest number of different geographic zones of the world, as well as the official HQ stations for each country's branch of the International Amateur Radio Union, or IARU. Now each country would count separately, and there were far more countries than there had been zones. As an example, zone 28 in Western Europe included more than two dozen countries: from Germany and Poland in the north, down to Italy, over to Greece, and then back up to Bulgaria and Romania—a total of twenty-nine separate political "entities," including tiny San Marino, the Vatican, and the strange-but-true Mount Athos, a remote Greek monastery that had been accorded political independence at some point in the past.

All in all, these changes would result in huge scores, since the 4,000 contacts multiplied by 400 or so different multipliers would produce somewhere in the vicinity of 5.5 million points. (Each contact was worth between two and five points, depending on the distance.) Every team knew the basics. Bottom line: Contact as many people (on other continents, since longer-distance contacts were worth more points) as quickly as possible while locating the hard-to-find stations tucked away in small countries and on islands. The best teams might beat that score—and they probably would. The team that scored well on their multipliers would be tough to beat if they were competitive on contacts.

The new rules were controversial. Doug Grant, the powerful chairman of the WRTC 2014, was a veteran of many international radiosporting contests, including several previous WRTC events. He believed that permitting only one station—one team member—to transmit a signal held back competition. "Why go through all that frustration," Grant explained, "when we could let both operators contact as many stations as they can? We decided to make it wide open."

But others disagreed. Many of the European teams that had participated in the Moscow WRTC 2010 preferred the "lockout rule," with only one signal at a time. For them, it emphasized the need for absolute teamwork. In addition, they felt that the new American rules would overemphasize the site location characteristics—specifically, the topography. Radio waves follow rules, traveling along the electromagnetic highways up to the ionosphere, where they bounce off molecules and electrons (these act as mirrors) and return to Earth. These pathways are variable, affected by the degree to which solar emissions energize the thin upper atmosphere. An optimum angle exists for the radio waves to "take off" from Earth's surface and "come back down" to the ground on the other end. These so-called propagation angles have been studied over many years and in different solar conditions. They vary with the frequency band and the height of the reflective mirror-like regions. In some circumstances, signal propagation across very long distances depends on multiple ionosphere-to-Earth bounces, or "hops."

During the twenty-four hours of this contest—or of any radiosporting event—solar conditions will vary, creating optimal and inferior angles of propagation (the way the mirrors reflect radio waves). The topographical shape of the terrain in all directions—specifically, the all-important direction toward Europe—could produce a different optimal takeoff angle. If a site were in a so-called sweet spot, its 100-watt signal would be stronger. From New England into Europe, one optimum angle exists at about 5 degrees and another at approximately 11 degrees. Nearly all sites would be equal, more or less, at the *higher* of the two angles, but they could differ by a significant amount at the *lower* angle. According to computer simulations, this difference could translate to as much as a ten-to-one effective power factor. In other words, a superior location could act as if its transmitter were ten times the power of an inferior spot at certain times of the day or during certain atmospheric conditions.

The Americans knew this. In fact, most of the international computer simulation software had been done by Hawaii-born Dean Straw, a brilliant man who had lived for much of his professional career in Connecticut and California. He had written extensively on what is called "terrain analysis." He also had consulted with the WRTC 2014 organizers, who went to almost superhuman lengths to locate the sixty-plus sites that were available—and that were suitable, according to both simulated signal strengths and actual testing of these signals as received in Europe. Working with Straw's program and the site selection team, Rich Assarabowski had developed a figure of merit (FOM) for each site. A tiny number of the sites actually were "too good," and several fell off the negative end, both simulated as well as measured in practice.

The Europeans in particular felt that any variances in the terrain at the sites, any differences that affected the optimum propagation angles to Europe—either on the FOM high side or low side—would be amplified by the new rules allowing both operators to go hell-bent-for-leather. They strongly preferred the Moscow competition's virtually identical grid-like arrangements of locating operating sites in huge flat areas, plus the one-transmitter-per-team "interlock" system that optimized teamwork, and also lessened any effect of the locations among these elite competitors.

The team that had finished second in Moscow was from Estonia. This team's leader had analyzed the new rule changes and realized that the key to this WRTC was going to be Europe. Everyone knew that, of course, since Europe had a huge number of amateur radio operators, especially ones skilled in Morse, and it was also a hotbed of radiosporting. An additional key would be the large number of different country-multipliers. Any site that had an advantage in getting signals into Europe would be in great shape, and likewise, any team having trouble reaching Europe would quickly fall behind its competitors.

Tonno Vahk, the Estonian qualifier and team leader, was a well-known European competitor and undoubtedly one of the top radiosporting operators. He had been an amateur radio enthusiast since the age of 13, following in the footsteps of his late father. His schooling, in the then–Estonian Republic of the USSR, was, in his words, "English-language-biased," in addition to his studies in Russian and his native Estonian. (By 2014, he had demonstrated that he spoke and wrote at the level of an advanced English scholar.) At 36, he started contesting from one of the club stations in Estonia, and over the past decade had built a major contest station of his own, where he usually operated by himself. He was a financial investment manager and, in his own words, had been "putting out some fires in our investments in Ukraine and elsewhere in the region. Bit of a headache, as you might guess." Physically he resembled Vladimir Putin, with thin, straight blond hair and an opaque but penetrating look that suggested he was the smartest guy in the room. Like many others at the event, he didn't appear to lack for self-confidence.

His teammate, also an Estonian, was 49-year-old Toivo Hallikivi, a cellular phone network marketing professional who was competing in his fourth consecutive WRTC. Their site was 5B, at Twin Valley East, part of the cluster of eight locations near the New Hampshire border. According to Vahk, their setup went smoothly, yet Vahk was concerned with the topography since there was a slight (ten-foot) rise toward Europe within the first 500 feet. A drop of forty feet then followed in the next quarter mile. A mile out in the general direction of Europe, there was a fairly sharp rise of nearly one hundred feet. Who could say what the effect would be?

⊳꙰ℓℓ╫

Steve London was completely unconcerned with the topography. Rail thin—he looks like he could jump up from his radio setup and run ten miles over the backcountry of New Mexico at any time— this super-focused former Internet network engineer had been prepared to skip the 2014 event after competing in five consecutive championships. He had agreed to join a team only when Kevin Stockton, his fellow competitor from the American Southwest, called. Like many others, at least west of New England, Stockton might have considered London the best single-operator contester in the United States, since London routinely placed in the top ten in national radiosporting events and often was the top dog west of the Mississippi River. He had medaled (bronze) in the 1996 San Francisco Bay WRTC, and had achieved solid mid-range finishes in each of the others. Their site, 2A, was called Kimball West and was located near the Estonians' site. London's primary concern was being too close to other teams, which could cause a sort of signal overload. When he saw that his site included only one other team, which was nearly three-quarters of a mile away, he relaxed.

London and Stockton went through their equipment setup quickly and without complication. They were ready to jump into the car with their driver and referee and return to the hotel for some rest when the head of site selection and one of the five judges drove up. Greetings and good-natured remembrances of past WRTCs and other contests were exchanged. Then the chief site man expressed concern that the beam team hadn't oriented the antennas properly—that is, toward Europe for the big aluminum antenna and the 7-MHz wire dipole. He thought the wire dipole was offset by as much as 35 degrees and insisted that they move the ends around. London and Stockton simply watched as the visitors went about the changes until they were satisfied that the team had the same "settings" as the other competitors.

Arguably, these four teams would have been on nearly everyone's "top ten contender" list. Each duo most likely felt they had a good chance to win. Soon, they would be head-to-head. Soon, we would know.

L to R: Dan Craig (N6MJ) and Chris Hurlbut (KL9A). Photograph: Bob Wilson (N6TV).

Latitude 13 Team, L to R: Tom Georgens (W2SC) and Dave Mueller (N2NL) with referee Chris Burger (ZS6EZ). Photograph: Bob Wilson (N6TV).

L to R: Toivo Hallikivi (ES2RR), referee Scott Robbins (W4PA), and Tonno Vahk (ES5TV). Photograph: Bob Wilson (N6TV).

L to R: Steve London (N2IC) and Kevin Stockton (N5DX) at their operating position. Photograph: Bob Wilson (N6TV).

*Chapter 7*

# MEANWHILE, AT OTHER SITES

A mile or so from the Myles Standish State Forest headquarters, you begin to feel tingles as you sense the depth of the woods. A solitary "WRTC site" sign had been hammered into the leafy ground by the highway, where it signaled the location of the Hawaiian team. The team qualifier, Mike Gibson, left no doubt that he was a quintessential rough-and-ready guy, with an unkempt Prince Valiant–style shock of straight, silver-gray hair and a matching mustache and goatee. His background was in engineering, and he had coupled that technical ability with practical experience building amateur antennas and supporting structures in Hawaii. Now working as a contractor to the US government, and more fit than men half his age, he built towers and their concomitant cameras and antennas along the US–Mexico border in the Southwest. Gibson had been a ham radio enthusiast since the age of 12 and a world-class contester for decades. He had competed at the WRTC in Brazil and finished well, just outside the top ten.

His partner, John Hillyer, had retired from the US Air Force as a jet jockey. Hillyer now worked as a commercial cargo pilot while maintaining his commission in the air force reserve. A regular jogger, Hillyer at 49 looked like he could pass any physical, as well as win a Tom Cruise lookalike contest. (Back in his *Top Gun* days, he must have cut a dashing figure.) Today he was a family man, married to a former anchor at one of Honolulu's major TV news channels. He also was building his own contest station in Hawaii.

Gibson and Hillyer's site, 15H, the first inside the state forest, was located in hilly terrain. After their smooth setup, Hillyer went out for a little run. As he said later, "I reached a slight hillcrest between our place and the direction of Europe, then looked back. Our antenna appeared below the height of the high point where I stood. We felt we were down quite a bit toward Europe."

⊳⟋⟋⟋⊦⫶

The Russian defending champions also were concerned about their location, officially listed as Airport East. There was no issue with terrain, since the location was quite level. However, Vlad Aksenov and Alexey Mikhailov found themselves only fifty feet away from a small taxiway at the Mansfield Municipal Airport. (Certainly there would be no problem with a too-close tree line.) They did share the airport site with one other team, which was located at the opposite end of the single runway. Yet there was private aircraft activity in the day, much of it training flights with associated noise.

This site seemed a bit of an insult to them—after all, they had won the Moscow WRTC in 2010—so they called headquarters and complained. To beat it all, now there was a *helicopter* hovering over the main runway! It wasn't directly over their tent, but certainly was a distraction during this initial visit to the site. It was Friday. Would this continue all weekend?

Organizing Committee Chairman Grant called the other team, located 300 feet away from the other end of the runway, and asked about the noise. "Do you have any noise from airplanes?" The team's referee answered that they didn't. "The Russians at the opposite end of the airport are complaining," Grant continued.

So, Grant offered the defending champs a fully qualified alternate site, one not far away. They made a quick trip to inspect it, but rejected it based on nearby tall trees. They felt the trees might interfere with the signals. Finally, and probably to the satisfaction of neither team member, it was decided that the site managers would construct a plywood wall to deflect sound from the taxiway. In addition, assurances were obtained that the helicopter would go away. The Russian team accepted the site—albeit grudgingly.

To make matters worse, the Russian team had problems setting up their station. One of their radios was a brand-new high-end model, which they had not used before and so were not intimately familiar with. A network problem developed, and their referee, Denis Pochuev, spent an hour trying to get everything working. In addition, the team called in the chief WRTC technical expert for assistance. The team had not brought a transformer to convert their 220-volt-based equipment to the American 110-volt standard; in addition, they lacked an interface to send Morse code from one of their transmitters. After a hurried series of both international and domestic phone calls, solutions to missing pieces of the stations were found. The essential keyer-interface circuit was located with assistance from a European who had an extra unit. The chief judge, Dave Sumner, actually drove out to their site with the voltage transformer. Now the station passed all the setup tests. Finally, after all this, it was getting late.

⯈〰〰┼╢

George DeMontrond and his partner, John Crovelli, were among nine WRTC teams located at Fort Devens, a large former military base just off Interstate 495, approximately twenty miles north of the WRTC headquarters hotel. The base had been closed in 1996.

A portion, including a large, unused former military airport, had been redeveloped into a commercial and industrial area. In addition, a significant area, 5,000 acres, now served as an active training ground used by reservist units of the US military as well as regional law enforcement agencies. Because of its history as a formerly active military installation, the WRTC organizers had been required to sign an unusual waiver that included the following stipulations:

WRTC 2014 acknowledges and agrees that UXO (unexploded ordnance) have in fact been identified at Devens subsequent to the Army's (previous) testing, and

Neither WRTC 2014 nor its employees, agents, or contractors will touch or otherwise disturb UXO or suspected UXO.

That document was part of quite a backstory on this waiver liability, and everyone involved in any way with the WRTC 2014 had to sign it. The areas set aside for the WRTC operating sites were clearly marked, and special limits were imposed.

DeMontrond and Crovelli rolled up to site 6G (Devens Davao Circle) in DeMontrond's large SUV. One of DeMontrond's longtime trusted lieutenants, who also looked after things at the ranch contest station outside Houston, had driven the vehicle with all its radio-sporting contents. They were pleased that the location was only thirty-five minutes from the headquarters hotel, since it would be easy to return if anything were missing. The area seemed a bit elevated—that was a positive—with very flat topography. They couldn't see much of anything, other than the Red Tail Golf Club that appeared to be not too far removed! That was fine with the big Texan.

Actually, DeMontrond was very serious about preparation. A driven type-A personality, he did nothing halfway. Doug Grant had been after him for quite a while to sponsor a team. Not only was that an important cog in the financial machinery of the championship, the Texan had also presented papers describing the bigger-than-life planning and construction of his ranch superstation at Dayton's annual ham radio get-together. He was a fearsome voice operator who often won when he dedicated a weekend to a contest. Grant had nearly threatened to hold an auction for the final sponsored slot until DeMontrond snapped it up. His was the last team in, which explains the (not-so-creative) name of "Team 59," which also corresponded to the signal report of a very strong station. They were thinking loud-and-big already.

On Crovelli's recommendation, DeMontrond had practiced in the "low power" (100-watt) category in a popular voice contest, finishing number 2 in the entire country to a perennial top operator in New England—a more favorable location. It was a most credible finish, to say the least.

The two men decided to do an actual dry run, using their complex WRTC setup, in the American Field Day exercise. They gave it a go from a tent on DeMontrond's ranch, using smaller antennas and low power; the permanent ham radio station was unavailable to them, having undergone a decade of metamorphoses—including near destruction following hurricanes and other nasty Texas weather.

It was late June, with 100-degree heat, threatening thunderstorms, and squadrons of mosquitoes and other flying peril. Their WRTC-compatible two-station setup netted 3,000 contacts, with two-thirds on voice and one-third on Morse code. With the advent of the Skimmer decoder software and the Internet, Crovelli knew that the WRTC ratio would flip heavily to Morse. The teammates talked about optimizing voice, DeMontrond's forte, to win the special award for the highest "phone" result at the WRTC, but decided against it. "We'll go up against the big boys," DeMontrond said.

All day Friday, the fifty-nine teams focused on setting up at their sites. The "show" would start at exactly 8:00 Saturday morning, so everything had to be in perfect operating shape when they left their sites Friday night. Their site managers would remain behind to watch over the equipment.

Like the other contestants, Ken Low and Scott Redd spent the day preparing. And like DeMontrond and Crovelli, they too had drawn a Fort Devens site, designated 6B: Devens Solarno South.

For Low, who was the qualifier, the WRTC was the culmination of a long and difficult trail. He had won the spot from an extremely competitive area: Maryland, Pennsylvania, and Delaware. This was the home base of two of the three most active and powerful radio-sporting clubs in the eastern United States: the Washington-based Potomac Valley Radio Club (PVRC) and the Philadelphia-centered Frankford Radio Club (FRC). Low had received plenty of help from his fellow PVRC members. During the three-year "sprint to the finish," he operated from four different competition-grade stations owned by members. He also joined teams outside the area. There was a spirit of cooperation and fellowship throughout—in fact, several station owners had been looking for operators and were ecstatic to get a competent contester to join their groups for a big weekend on the air. At the end of the process, Low emerged with the highest qualifying score. He had worked hard, gained access to great stations, and gotten a few breaks. Here he was now, with a terrific partner at his side.

At 69, Scott Redd was older but, as you might expect from someone with his military background, driven and disciplined. He was adept at using laptop computers and their software programs. Low and Redd had practiced in late May, only two months earlier, in a worldwide Morse code contest. They used modest antennas similar to those in the WRTC systems, along with the same computer and networking systems they would be using in New England. They operated from the Washington, DC, area. Everything went well. In addition, Redd had been operating for several years from an authentic superstar contest station in New Hampshire, and he was familiar with the fantastic conditions to Europe from that area. People joked—and sometimes it seemed to be true—that someone could climb to the top of a roof (or even better, to the top of a tower at a ham radio station) and *see* Europe. It was right over there, if you looked hard enough. Radio waves seemed to agree, since New England conditions were the envy of the entire country.

When the team arrived at Devens, along with their Serbian referee, everything started up smoothly. Even though Redd had been battling a bad respiratory infection for two weeks, he soldiered on and the station came together without a hitch. It seemed easy—if working in a very warm and stuffy tent could be described that way. The big aluminum rotatable antenna worked fine, as did the wire dipole for 7 MHz.

At 4 p.m., they were nearly ready to drive back to the headquarters hotel for some rest. The final test, which seemed trivial, was to transmit into the other wire dipole for 3.5 MHz. But as soon as the antenna was connected, the receiver nearly exploded with a hideous electrical noise. They were crushed. What in the world could be happening? The antenna *transmitted* okay, but it was totally useless because they could *hear* nothing but loud electrical hash. They considered calling headquarters and requesting an alternate location. But at nearly 5 p.m., the thought of disconnecting the equipment in the hot tent, driving somewhere else, and doing everything all over again seemed too daunting.

Two hours of intense troubleshooting followed. Low and Redd tried every combination they could think of. The site manager tried to help and called George Wagner, a volunteer from Florida who was at the neighboring DeMontrand-Crovelli site. Through a process of elimination, using Wagner's mobile high-frequency setup in his car, they discovered that without the generator, there was no noise. But why was the generator producing such horrible interference, and on only one of the five frequency bands? By now it was close to 7 p.m. The sun was getting low in the summer sky when the site manager asked, "Is the generator grounded?" Apparently, grounding the generator either wasn't part of the standard site preparation, or else it had been missed. They moved the Honda unit a hundred feet back into the woods. One of the ground stakes from the tower was removed and connected to the generator. They all held their breath as it was restarted. *Voilà!* No noise at all.

They were home free—or so it seemed. They ran the station through some final paces on all the bands. Everything looked great, and they were ready to switch off the gear and leave the site for the evening when their cell phone rang. A Brazilian team, another of their neighbors at Devens, reported that Low and Redd were generating a trashy signal, described as "phase noise," across the entire 14-MHz band. The two had no idea how that could happen. Both their radios had reputations for very clean signals. By this time it was nearly dark. In a series of somewhat emotional and heated cell phone conversations, with no one at the Brazilian site fluent in English and no one at the Low-Redd site able to speak Portuguese, a palaver somehow took place. Both groups would transmit to a neutral site, one that was equidistant from both teams. Once again, they turned to their neighbors at Devens, the DeMontrond–Crovelli team. (Fortunately these two contestants weren't over at the golf club mixing with the locals!) The suspect radios were declared by the rarified ears of John Crovelli to be clean—not at fault. The international compromise recognized that, at 500 meters apart, they simply were very close, and when beaming toward Europe, which was most of the time, the signal was too strong. Neither station would need to move to a backup site—it probably was too late for that anyway. Both teams would just have to live with the reality of the situation. They had to stay far enough away from each other in terms of frequency separation on each band. Finally, at 8:30 on Friday night, Ken Low and Scott Redd left for the hotel. It had been a long day.

▷꧁꧂◁

The Lithuanian team's setup on Friday was also a nightmare. Its origin apparently stemmed from team leader Gedas Lucinskas's simple attempt to charge one of their mobile phones before attending the opening ceremonies on Thursday. At the time, as he disconnected his Lenovo laptop, he noticed a "spark" of some sort in the 220V/110V adapter. Immediately the PC went into "hang-up" mode: unresponsive to keyboard, mouse, or "control-alt-delete" command. The next morning, during the station and site draw mega-meeting, a message came in from their friend, a third Lithuanian who had traveled with them to provide technical support. The PC wouldn't boot up at all: It displayed only the dreaded "blue screen." Lucinskas, a telecommunications engineer, had over forty years' worth of top-level radiosporting, including previous WRTC battles in Moscow and Slovenia. On the way to Boston, he had recalled an old Lithuanian adage, "The third time never lies," which is a version of the English "Third time is the charm," and thought that 2014 might be his lucky world championship. Yet this wasn't a good start, for sure. He and his teammates stopped at the hotel's Information Technology (IT) support center and inquired about buying a new PC. The people there kindly offered to loan them one of the two Dell laptops on hand. The problem was solved, it appeared.

On the way out to their site on Friday, the team's technical guru tried everything possible to bring the previously trustworthy Lenovo PC back to life, but with no response. They stopped at a Radio Shack and bought an external keyboard and USB cable so they could load all the contest-specific software onto the borrowed Dell. Their site was a good sixty minutes from the hotel and was one of three Freetown locations—with additional descriptives such as "Hill, Rock, and Knoll," thus implying some elevation. They connected the station, and the borrowed computer booted up and appeared to work properly. The station seemed to "hear" and "get out" well. Lucinskas and Jukna believed they had overcome the dead-PC problem. It had certainly been a long, stressful day. The competitors, referee, and mobile technical support man, along with their driver, got back to the hotel around midnight. They planned to grab four or five hours of sleep and be ready for a good start in the morning.

Mike Gibson (foreground) (KH6ND) and John Hillyer (KH6SH). Photograph: Bob Wilson (N6TV).

George DeMontrond (NR5M) and John Crovelli (W2GD) at their operating position. Photograph: Bob Wilson (N6TV).

L to R: Ken Low (KE3X), referee Dusan Ceha (YU1EA), and Scott Redd (K0DQ). Photograph: Bob Wilson (N6TV).

# Chapter 8

# ONE FINAL OBSTACLE

~~~wwv~~~

Back in the far reaches of the Myles Standish State Forest, in the wilds near Camp Squanto, things had gone smoothly at site 15W. In fact, the setup had gone so well that everyone felt relaxed. Manfred Wolf and Stefan von Baltz had connected their equipment and, other than some strong signals from their radio neighbors—this was expected—they were all set to go by early afternoon. In addition, at the end of the rough forest trail where they were located in the southwestern corner of a seven-team cluster, they had the advantage of being on the "back side" every time another team beamed toward Europe, which was most of the time.

Now only one thing remained. The Score Collection Computer, or SCC, needed to be connected. The SCC was the real-time score reporting system that compiled the numerical score, as well as details like the number of contacts and multipliers. Previous WRTC organizers had recognized the desirability of a scoreboard during the contest. After all, this championship attracted spectators. Although those spectators might not be sitting in bleachers or stadium seats, they did follow the results, and they followed them from all over the world. The competition was exciting, and the competitors and referees had traveled great distances to participate. Beyond that, friends, work colleagues, family, and fellow ham radio enthusiasts were eager to see how individual teams were doing.

At the Finland WRTC in 2002, Nokia cell phones had been used. (Many of the key organizers were executives of the company, based in the Helsinki area.) The phones were issued to each referee, who texted his team's score to the collection system every hour. These scores were compiled, ranked, and posted worldwide on the Internet, which was a significant development and a positive step.

For this year's WRTC 2014 in New England, a unique system had been developed to access the data from whichever software program a team was using, compile it in the SCC, and send it back over the Verizon network to a central SCC hub that was connected to the Internet. The system worked fine *if* the sites could connect on the cellular network. But only about a third of the sites connected all of the time. The organizers knew in advance that the location of the teams could be a weak link, since cell phone companies didn't prioritize coverage in the middle of proverbial nowhere—and that's exactly where many teams were.

Wolf and von Baltz were relieved that their amateur radio stations worked flawlessly. They began to contact other hams using their German call signs with a "portable W1" designation, which showed they were in the New England region of the United States. Soon both guys were "running" both North American and European stations lickety-split. They delegated the job of connecting to the SCC to their referee, Wes Kosinski, who recruited me to assist. The instructions included registering site 15W on an older cell phone that was included as part of the setup instructions. The dated gadget appeared to be multiple generations old. To begin with, neither of us—so-called wireless wizards—could figure out how to *turn on* the little thing. Was the battery dead? No, because we removed it and measured the voltage on a voltmeter—4.1 volts, perfectly acceptable. Even more frustrating, we couldn't find the "power on" switch. Having failed this basic IQ test, we then discovered that the phone was an early clamshell model. Opening it produced a button labeled "PWR-END." The two words seemed contradictory . . . yet pushing the button brought the little critter to life. Aha! Step one.

The next step was supposed to send a text message. After trying several combinations of buttons, at some point the screen—and it was teensy—displayed "Message." Somehow in the process, another window popped up and we typed, on microscopically small keys, the designated ten-digit phone number for our SCC masters back in civilization. Now on a roll, we pushed "send." Immediately a new screen opened with a text box. According to the instructions, we were supposed to type in "my team." We did and immediately got the response "Your team is DJ5MW/DL1IAO."

Okay, now we were cooking. But another problem arose. Even though our *phone* appeared to be registered, the *modem* was supposed to show a green light to confirm proper connection to the "mother SCC" back at headquarters. Hmm. No "greenie." Nothing at all.

All this time we had been tiptoeing around the other wires, cables, and connections inside the sweltering tent, trying not to disturb the two competitors, who were testing out their radios, quite busy with what appeared to be a huge string of callers. Certainly this was a positive sign that the site "worked."

Wes and I sweated inside the tent while Manfred and Stefan (all of us were now on a first-name basis) declared the radio gear *"Alles in Ordnung"* and retreated to cooler environs outside by the campsite. They were ready to go, and it was our responsibility to get the score computer ready for action.

Back to the instructions. Our situation was actually described, and then we were told (more or less) the following: In the case of marginal cell phone conductivity, you may need to use the additional antenna to improve signal strength.

Wes rummaged around in the small SCC container bag he had been provided at the hotel. There it was: a single piece of wire approximately five feet long. That was it? The magic antenna?

Apparently, it was. The magic antenna was five feet of thin black insulated wire with a minuscule thread of stranded copper within. We placed the so-called signal-improver at every conceivable position over the operating table. Along the back of the tent, the Germans had hung a large banner from their radiosporting club back home: Bavarian Contest Club. Apparently, neither the BCC banner nor the SCC antenna doodad added any gain. After a half hour of experimenting with all possible positions, I removed the "booster antenna." So far, it had no boost. It looked as if Wes would have to telephone the score into headquarters every thirty minutes—and that certainly was *not* a pleasant prospect.

The instructions also mentioned trying additional positions: *The SCC connection might be location sensitive.* I figured if the cell phone coverage with Verizon was marginal—and it certainly was, out here in the boonies—then why not use our own cell phones as an aid? Since my own smartphone also was on Verizon, I oriented it around. In one direction, the signal strength popped up from zero to one measly little bar. At least we knew where the base station was located. The signal was faint, but it was (sort of) there. I moved the small SCC a quarter inch and pointed it in the direction of the single Verizon strength indicator. Immediately the modem's green LED came on and began to blink!

Wes and I emerged from the tent and announced our breakthrough to Manfred and Stefan, who were relaxing one hundred feet away with Mike and Derek Bennett. The Bennetts were the perfect site managers. The elder Bennett was in the well-earned sunset phase of a career in administration. A man with a reserved nature, he didn't say a whole lot, but what he said was to the point. His son, Derek, was more outgoing—garrulous, in fact. He had a bachelor's degree in political science and a job in nonprofit advocacy. Derek had a somewhat typical background in amateur radio, an interest that had started when his Boy Scout troop leader brought a ham radio set to camp. Following that introduction, Derek bought an inexpensive shortwave receiver and had fun doing what he called "tuning roulette," putting the radio on the "scan" mode until it stopped on a program. The scanning process would end automatically once the circuitry captured a loud-enough signal. In this way, he would happen upon broadcasts such as the "Voice of the Andes," HCJB, in Ecuador; the booming signals from the "Voice of America," BBC; and other big international short-wave broadcasters.

Father and son later took a class in amateur radio together and got their licenses at the same time. Their call signs were almost identical, differing only by the last letter. At a meeting of their local radio club, a representative from the big club in New England called the Yankee Clipper Contest Club (YCCC) made a presentation about the WRTC 2014 and asked for volunteers to be site managers. While this assignment might be daunting to most people, the Bennetts found it exciting. They could camp out, provide security at one of the locations, and watch the beam team come in and erect a tower and antenna system. Best of all, they would get to meet and see up-front-and-personal one of those (as they were described in the program) world-famous radiosporting teams.

Another member of the local radio club, and part of the site team, was a perky woman named Judy Attaya-Harris. She, along with her son, somehow had managed to navigate the forest trail to find the clearing reserved for bird hunting, which was now the radio site location. Anxious to learn everything she could about her new hobby, she was chatting with Manfred, Stefan, and the Bennetts. Both she and her son, Anthony, were relatively new to amateur radio, but her enthusiasm was infectious. She announced that she had volunteered to make a food run for the team. When she learned that the competitors and referee were from Germany and Poland, she pointed out that there was an Aldi store in the area. (The area? You had to go nearly ten miles simply to get out of the forest!) Judy proceeded to take orders for lunch as well as the rest of the weekend's meals, and soon she was on her way to the German-owned emporium to buy a cornucopia of European good-ies: Black Forest ham, sausages, unique German cheeses, and dark breads. Of course, our three visitors were elated. With the strong outdoor woodsmen experience of Michael and Derek, plus Judy's outsized exuberance and interest in "this whole WRTC thing," site 15W looked to be in very good hands.

One of the amenities at the Bennetts' campsite was a radio, which was kept tuned to a local station, WATD, on 95.9 FM. The announcer often referred to the WRTC 2014 and mentioned bits of information between songs—a steady stream of positive pro-motions. The Bennetts filled us in on the background of the radio station. A local person had applied for the new station license and construction permits, and then he presented his plans for the small building and associated tall tower to the local planning commission. A wave of objections soon erupted from landowners and residents who didn't want the tower to be nearby. So, a Plan B went into effect: A site near the town dump was available—for some strange reason, the site hadn't sold previously—and this time around there were no objections. The relieved station owner decided that the call letters would represent the process, and so his new FM station became WATD—for *We're at the dump.*

The plan was now clear. Soon everyone could enjoy the food and relax. I would pilot the Dodge Charger back to the hotel, and Manfred and Stefan could have some downtime before the very early getaway Saturday morning and then the grueling, uninter-rupted stretch of the contest itself. Only one thing remained: We needed to try out the simple arrangement to provide headphone audio to Wes, the referee, who would be seated in the middle with Manfred on his left and Stefan on his right.

Stefan had brought a small metal gadget for that purpose. "I got this from a friend of mine. We tried it and it worked just fine at home," he said as he reached into a large duffel bag and brought out the *Kopfhörer Verstärker*, as it was labeled in German, or "Headphone Amplifier" in English. The connections were easy: one audio cable to each of the two radios, and another for Wes's headphones. What could be simpler? Stefan plugged in the cables to his station and then to Manfred's setup. They both listened and indicated that the audio was fine. No problem.

Wes sat down and plugged his headphones into the *Verstärker*. "I do not hear anything," he said slowly in his slightly accented English. "Nothing at all."

Every combination was tried. Each connection and every plug-in was checked. Headphones were swapped and exchanged. Cables back to the two radios were replaced. When Manfred and Stefan could hear their own radios properly, Wes's radio had no sound at all. When Wes could hear, the two of them couldn't. In one strange situation, Wes could hear, sort of, but then Manfred was listening to Stefan's radio and vice versa. Chaos seemed to be the rule.

Judy's return from civilization was perfectly timed—and for me, miraculous. She seemed absolutely comfortable with navigating the state forest. By contrast, our original difficulty finding our way through the labyrinth of roads to the site itself, plus the forest-path access off Cuttersfield Road, had convinced me that the beam team members who had installed the tower, antennas, and tent, and "Daniel Boone" Bennett and his son, would be the only other human beings brave enough (or GPS-enabled) to find our little isolated piece of heaven. Our supercharged Dodge Charger certainly hadn't been the ideal off-road car.

We needed a break, and Judy's positivity was a perfect tonic. So was the food, a veritable table-of-plenty that soon overflowed the small camp table: European goodies, super-crunchy peanut butter, crackers, several varieties of cookies, orange juice and soda pop, plus water. Several large chests full of tinkling ice cubes stored the drinks and all the perishables.

Manfred, Stefan, and Wes took a break from exchanging cables and swapping headphones, the frustration of the past hour dissipated, and everyone joined together outside to eat. It was a good break for fresh air since the tent was hot and musty. Yet thirty minutes later the team and I were back inside to sort things out. Manfred and Stefan were trying yet another set of headphones to see if any combination would work.

Just as the same old things were acting like the same old things once again, we heard a car creeping across the clearing. I recognized the occupants as the main players in the judging process for the entire event. Dave Sumner, widely known in ham radio circles by his call sign, K1ZZ, was the driver. He had served as a judge (or chief judge) at all WRTC events since 2000—he would be chief judge for this Olympic-style competition—and within both the United States and international amateur radio circles he was a true VIP. Besides volunteering for the WRTC 2014, his "day job" was chief executive officer of the American Radio Relay League, or ARRL. The ARRL is the national association for amateur radio, with 163,000 members in the United States and Canada (as well as other countries). A century old, the "League" is the primary representative organization of 700,000 amateur radio licensees to the US government. If that weren't enough to occupy Sumner, he had served as secretary of the International Amateur Radio Union, the federation of all the individual national groups. He has said that his degree in political science was augmented (or diluted) by numerous hours at the Michigan State University radio club station on campus. Since that time, Sumner had worked for the ARRL continuously, and was considered a topflight radiosporting operator on his own.

Sumner's site tour team included the two men who would be key technical contributors to the judging committee. Between them, they had created much of the software and database-analysis tools to crunch through huge amounts of logbook data sent in electronically following a radiosporting event. There would be only twenty-four hours or so from the end of the WRTC at 8 a.m. Sunday until the final ranking was decided. The overall order of finish was important to each of the teams. However, the prime responsibility of the judges was to get the *top* teams in the correct order, and especially to get the top three—the medalists—right. These two technical wizards had years of experience, both as elite contesters and as experts in software and data analysis.

Larry Tyree, who went by "Tree," lived near Portland, Oregon, and was known in the radio world by his call sign, N6TR. Trey Garlough, N5KO, hailed from a small ocean-side community near Santa Cruz, California. They were "the guys" in terms of the technical aspects of data analysis. Tyree, especially, had developed a sixth sense for anything that even hinted of shenanigans. The eagle-eyed data geeks were hoping to let computers do the work while the two of them stood by with little to do but print out the results. Now, on a sunny Friday afternoon, they were simply making the rounds with Dave Sumner and relaxing.

The visitors introduced themselves to the Bennetts and Judy Attaya-Harris. They were well aware of the importance of the volunteers. Without them, regardless of the financial support and worldwide interest in the amateur radio community, this event wouldn't be possible. Sumner, a slight, balding man wearing wire-rimmed glasses, had a shallow voice and reserved speaking manner that belied his responsibility. Tyree, on the other hand, looked like the former Intel man and West Coast high-tech software expert he was, with longish hair surrounding a prematurely balding pate. He presented a completely different aura than Garlough, who had worked for a Santa Cruz area start-up and had benefited from its acquisition by Cisco Systems. Garlough was quiet and unassuming, the very personification of mellow. Both knew their way around database systems—that was certain. With a nod to our food-laden table, I led them over to the tent where Manfred, Stefan, and Wes continued to wrestle with the *Kopfhörer Verstärker*, with little to show so far except exasperation.

The problem with the team's audio network was explained. Did they have any ideas? Perhaps the headphones themselves had something to do with it, Wes suggested, since exchanging them with some backup headsets seemed to make things better, or worse. Tyree, with the strongest and most current technical background, worked with the setup and listened to several of the headphones. He shrugged and said, "Keep at it. You'll get it." Hmmm. Not as promising as we'd hoped.

"We could use another pair of headphones. They might be the problem," said Wes, partly as a joke and partly serious.

With that the three visitors left the tent and started back to the car, where they were intercepted by Judy. Sumner paused, and she asked him, "Mr. Sumner, what is your role here in this event?"

"Call me Dave, Judy. I'm one of the judges."

I stepped in. "He's the chief judge, Judy."

"No. Really, there are five of us. We all have an identical responsibility," Dave said.

"Oh. I see," she said with little conviction. "Do you live in this area?"

"No. In Connecticut. I work for the ARRL," he answered.

"What do you do for the AR . . . RL?" She stumbled with the acronym. That led to an honest round of muffled laughs as she queried the head of a rather large organization. "Did I do something wrong?" she asked.

"No, no," I said. "Not at all. What *do* you do, Dave?" I prompted. All the while, Judy held a small notebook and was furiously taking notes.

"I do a little of everything," he said.

"He's the CEO, Judy," I said. "He's the top guy at the League."

"Oh! I'm so embarrassed," she said.

Sumner was comforting. "No problem at all." He glanced from Judy to the Bennetts. "Thank you, all of you, for your support. Now, we'd better be off. We're trying to visit as many of the sites as we can today," he said.

Judy whispered to me, "I had no idea who he was. He seemed so nice. So reserved."

With a final round of handshakes and a quick pass through the appealing cookie selection, the trio waved good-bye, got in their car, and began to turn back in the direction of Cuttersfield Road. Suddenly the car stopped and started backward. Tyree bounded out and handed a box to me. "Brand-new headphones. Dave Sumner's. He said to try them."

It pays to have contacts in high places.

Yet the Black Forest ham and other Teutonic specialties didn't alleviate the sense of foreboding. One thing seemed clear: There were what are called "ground loops" in the little box—otherwise, Manfred and Stefan couldn't have heard each other. In addition, every time Wes had tried to listen, the audio to the two Germans was cut off, or nearly so. What about grounding? Perhaps there was some sort of ground problem. Maybe if the two radios were connected more firmly to the earthen ground? Everyone went outside to check. It was resolved to remove one of the tower grounding stakes and put it right next to the tent. We would cut off a short length of wire from the generator's ground system and use that to connect the new station ground near the tent to the equipment. Maybe that was the problem. We tried, but nothing improved the situation in the least. If Wes wasn't able to listen to both operators' audio streams and hear signals transmitted *and* received, he couldn't perform his duties as referee. The team wouldn't be able to compete!

It was nearly 8:30 p.m. The sun had set behind the tree line, and a long dusk was nearing its end. Darkness was closing in when

Stefan reached into his duffel bag once again. "I have some ideas," he said in his accented, slightly sibilated English. After a moment of furiously searching, using only the small light on the equipment table, he produced two dark gray items. Each one was cube shaped, measuring a couple of inches on the side. "Audio transformers," he added. "I think we need to eliminate this," he pointed to the *Verstärker*, "and make our own circuit." As three sets of eyes nervously watched from the door of the tent, Stefan stripped the ends from the wire and used his fingers to twist the metal strands to the connector post of one of the transformers. It was a primitive form of what is called wire wrapping, and it (more or less) made a connection. He fished around again and brought out an alligator clip, which he used to clamp the "wire wrap" to the metal post.

"Ah, that is better," he said.

Slowly a homemade audio-isolation circuit took shape in the dimming, tense silence. The little electronic device would allow his audio to reach Wes's headphones in good shape—if it did what it was supposed to do. But would it? Wes slipped into the tent. Both men listened. It worked! Stefan could hear his radio perfectly, and so could Wes.

"Let's solder it together," Stefan said, "and then we can build one for Manfred." With that, he turned once again to his Noah's ark of electronic goodies and brought out a German soldering iron that appeared to be made to last at least a thousand years. He plugged it into the power strip, found a roll of solder, and sat back in his chair. Once the circuit was built for both guys, I could get them back to the hotel for some now desperately needed rest. The handheld little beast came to life, began to heat up, began to heat up more, and continued to begin to heat up. But it never became hotter than a simple hand-warmer.

"Oh my. It's a 220-volt iron," said Stefan, now recognizing that it would never get hot enough on the American 110-volt electrical system. With that realization, he took the two small audio transformers, some random wire, and some solder, and cradled them in his hand. We walked to the car in a now completely dark forest site, illuminated only by the table light in the operating tent plus a single lantern in the Bennetts' tent. Stefan needed to get back to the hotel, locate a 110-volt soldering iron, construct his system, and hope it would work. His shoulders drooped, and his face was a sheet of fatigue as he and Manfred got into the rear seat of the Charger. Wes rode up front. I started the engine, and the headlights projected a narrow beam of light into the clearing. We swung around and headed back along the narrow path toward what I hoped would be the road. It was nearly 9 p.m. The headquarters hotel was an hour and a half away.

L to R: Derek and Michael Bennett. Photograph: Wieslaw (Wes) Kosinski (SP4Z).

L to R: Anthony Attaya-Harris, Judy Attaya-Harris, Michael Bennett, Derek Bennett, Manfred Wolf, Wes Kosinski, Stefan von Baltz, and Jim George at site 15W. Photograph: Wieslaw (Wes) Kosinski (SP4Z).

L to R: Trey Garlough (N5KO), Dave Sumner (K1ZZ), "Tree" Tyree (N6TR), Manfred Wolf (DJ5MW), Stefan von Baltz (DL1IAO), and Wes Kosinski (SP4Z) at W1P. Photograph: Jim George (N3BB).

Chapter 9

# A BAD TRIP

As our Dodge Challenger inched along, the WRTC table light and the Bennetts' campsite lantern faded into a single point of light. The visible dot abruptly disappeared when I turned left at the edge of the clearing. We approached the dip into the wet-weather creek in complete darkness, other than the muted dashboard lighting and the headlight beams. We crossed slowly. What would we do if the car bottomed out on a rock? Over on the upslope now, bleached stones and caked mud suddenly changed to a level forest trail again, and the headlights formed small lighted slits into virtual nothingness. With the four of us inside, the car ran low to the ground. Limbs brushed against the doors and made grating noises as we passed, a reminder of our isolation.

After what seemed to be forever—perhaps 200 yards of allowing the car to seek its own speed with only the lightest pressure on the accelerator—we emerged at the WRTC site 15W sign. I paused when we reached the pavement of Cuttersfield Road. Several hundred yards away we could see the lights from the site of Tom Georgens and Dave Mueller. The mood in the car relaxed a bit now that the most difficult part of the trip was over.

But was it? I recalled the "directional disorientation" I had experienced finding the site. We had been on Circle Drive when the weather-beaten arrow and sign indicating "Camp Squanto" appeared. Cuttersfield Road, with its faded and battered tarmac, was anything but a main road. Now we found ourselves in inky darkness back at the same junction. I knew we had come in from the left, but there were large one-way arrows painted on the road.

It was 9 p.m. A sense of calm enveloped the three passengers, but I was nervous. I wouldn't admit that I had no clear idea how to get back to the main road, Fearing Pond. With no other choice, I turned the Charger to the right and drove at twenty miles an hour, hoping for some sort of road sign to indicate the way. We continued around a broad curve, which seemed to bear to the left. After a mile or so, an intersection lay ahead. No signs, as usual. We were still on the road marked with the one-way arrows. There were two options: one to the left, and one to the right. If I hadn't been so tired—or had learned how to use the GPS system on my smartphone—perhaps I could have reasoned that we had been on Circle Drive and had come full circle back to where it began, back on Fearing Pond. If I had done that, then a right-hand turn would have been logical and would have started us back to the state forest headquarters and the way out of the maze. But somehow, right then, I had an instinctive sense to go left. After all, we had just turned right. Didn't we need to go back at the first left?

I made the turn to the left, onto what was a wider road. Perhaps it was Fearing Pond Road. But mile after mile passed with no familiar indicators. Where was that abrupt turn where the bridge was out? On we went, until after what seemed like an eternity (but probably was only ten miles), we reached an intersection. Surely there would be some mention of the headquarters. In the dark, one small weathered road sign indicated a town with a name I didn't recognize—eight miles to the left—and a Massachusetts state highway with an unfamiliar number to the right. We were lost. Totally lost—now it was clear. It might have been the logical play at the time to go to the town of "Somewhere," Massachusetts. If someone was up and about, we could have gotten directions. Instead, for reasons not now apparent to me, I decided to return to the forest intersection deep in the woods and go the other way—go right, counterintuitively, at the cursed intersection. We did just that.

We backtracked. No one said a word. It was approaching 9:30 when we reached that intersection, or what I hoped was it. I went on straight. Straight into the oblivion, making effectively the right turn we hadn't taken before. Within a quarter mile one familiar landmark appeared, then another. Five minutes farther, the abrupt turn alongside the "bridge out" sign appeared. Relief! Several intersections later, the confirming godsend: "State Forest Head-quarters," with an arrow pointing to the left. To the left we went, and within five miles we had emerged from the woods alongside the HQ building. At that point, we turned sharply right and headed back onto Cranberry Road, where we passed the ebony nothing-ness of the bogs before reaching Tremont Street once more. The nightmare was over.

Stefan spoke for the first time since we left the site. "Please stop the car. I need to get out." From the indirect glimmer of the Char-ger's dashboard, I could see that he looked pale and nauseated. "Sorry, but I need some fresh air."

After all the twists and turns in complete darkness, mixed with the uncertainty of unraveling the defective audio box, he'd become sick. I pulled onto a gravel patch off the road, and slowed before stopping the car. Stefan got out and walked to the rear, out of view. With no other cars on the road, we waited in silence with the door open and the Charger's headlights piercing straight ahead into the night. Although I listened for sounds of sickness, nothing other than Stefan's footfalls on the gravel and the idling of the motor pro-truded into the muted evening. When he returned to the car, Wes moved to the rear seat and Stefan sat up front with me. This way it would be easier to watch him and pull over if necessary. My nervous driving with the quick stops and abrupt turns had taken its toll.

After a few more miles, and what appeared to me to be yet additional unfamiliar landmarks (I was frazzled by now), we finally reached a sign pointing to the right: "To 495." At last we could leave this century-old setting of bogs and back roads. Once on 495, with its interstate highway standards, the nervous tension softened. No one spoke for thirty minutes.

Finally, Stefan broke the silence. "Can we stop somewhere? Get something to eat?" he asked. "Perhaps some fast food. I don't want to have to wait long." He wasn't a small man, yet he spoke with a soft manner. As if on order (it was time to have something go according to plan), the next exit was coming up. The standard signs listed gasoline stations, motels, and restaurants. Perfect. Part of me hesitated to spend time over and above what would now be a very late-night return. But the frantic effort to provide the referee audio had been so focused that no one had eaten anything since early afternoon. We exited the freeway into a cluster of chain motels, two brightly lit gas stations, and some restaurants. The same familiar names dominated here as they did at nearly every interstate off-ramp across the country.

A minute later, we pulled into a McDonald's. At this time of night only a few cars were in the parking lot, and I pulled the Charger into a well-lighted spot near the door.

The fluorescent lighting inside felt jarring as a short young woman cheerfully welcomed us with the prescribed greeting. There was no rush to make a decision. We were the only people in line at that time of night. Stefan ordered a kids' Happy Meal, Manfred a burger of some sort, and Wes and I both asked for a large order of french fries. I picked up the tab—it wasn't much—since I felt responsible (*guilty* is a better term) for our meanderings in the labyrinth.

"And which toy would you like, sir?" the attendant asked Stefan. "I do not understand," he answered slowly, parsing his words. "Boy or girl?" she said.

He hesitated and smiled. "I think I am a boy," he responded, with a slight *sss* sound on the front end of his *th*. We all laughed when she handed him the small toy that completed the Happy Meal. We sat down at one of the empty booths and ate without doing a lot of talking. Stefan seemed better, but the audio problem, coupled with the inky labyrinth of the state forest, cast a pall over the late-night stop.

▷────

The rest of the trip should have been easy, since we had only thirty or forty more miles to go on 495. The clock indicated precisely 11:00, illuminated by the Charger's rather odd color scheme of red and black on the digital dashboard. Maybe a half hour more, then hopefully the guys could find someone, anyone, who could loan them a 110-volt soldering iron. They could finally try Stefan's idea of cobbling together a circuit that would blend in all three audio streams without messing them up so horribly.

The DoubleTree hotel was located in a complex New England cluster of roads linked by three major highways: Interstate 495, which forms a sweeping arc running from north to south around Boston to the west; the Mass Pike (Interstate 90), which bisects 495 at a perpendicular angle as it runs from Boston all the way to the Pacific; and venerable old Route 9, the original roadway connecting Boston with suburbs to the west, which runs pretty much parallel to the Mass Pike and has a gazillion traffic signals, intersections with stop signs, small-town centers, and strip malls. Final access to the headquarters hotel complex was tricky, since it was located on an access road reachable only from Mass Route 9. The hotel was in a typical Boston-area high-tech enclave, and it had been hard enough to locate in the middle of the day after the trip from Boston's Logan Airport. And now, exhausted and emotionally fragile, I needed to exit 495 and find the DoubleTree in this network of major highways and spiderweb of connecter roads with names like Computer Drive and Research Drive. The WRTC organizers, those modern masters of wireless hieroglyphics, had provided directions *to* the state forest in our package. But since no *return* directions were included, the masters must have assumed that if the teams managed to get from point A to point B, they would be able to retrace those steps all by themselves. But near midnight? After the sort of day all of us had experienced? On a dark night, where the "goes-into" was different from the "goes-outta," in terms of that hotel access? All of this flashed through my mind as we approached our target destination.

You might expect, actually you *would* expect, that a car full of shortwave radiosporting gladiators and their world-class support team would have printed out independent return directions. Or maybe a Google map of the complicated intersection at the DoubleTree. Wouldn't at least one of these wireless warriors pull out his smartphone and access the GPS app? In fact, that sort of planning or reacting should be demanded. But, truth be told, I had failed to try out the navigation feature on my new super-duper Motorola phone with all the cool features. I was the driver, the *American* driver, in *America's* New England. We were only a couple of miles away now. But somehow, some way, at 11:15 at night, none of these logical steps were taken.

A sign ahead indicated Exit 22. A voice from the rear of the car said, "Here it is. Take this exit." And despite knowing better, I reflexively did just that.

Ten seconds later, we moved into a huge sweeping loop, past any chance to pull over and go back. The roadway morphed into a monster-sized cloverleaf cut into solid rock hills. Ahead of us, a super-sized roadway sign announced that now we were entering the Massachusetts Turnpike, a toll road, and the next exit was eleven miles east in the direction toward Boston. I had made a very bad decision. We were beyond the point of no return. So, on we went. The car was sickeningly quiet now. No one reminded me of what a boneheaded driver I had become. I rocketed the Charger up to the speed limit, and the sterile silence was punctuated only by the highway sounds. I focused on getting us to the next exit and a recovery plan, Plan B, which now was a Plan C or D, given our trip so far.

A huge roadside service area suddenly came into view. It was a welcome development on two fronts: Hopefully, I could get some help with directions, and our little sojourn down to the state forest and back (well, almost back) had left the fuel gauge at about the one-quarter mark. So, we pulled around the eighteen-wheeler trucks into one of the many gas pumps and stopped.

The fueling station was nearly deserted at such a late hour, and, if it were possible, our already frustrating trip took another weird turn. I couldn't find the switch, or button, or release, or magic potion—whatever it took to open the gas cap on the Charger. I looked everywhere: the front panel, the glove box, under the instrument panel. Where in the world did the automotive geniuses put that control? Stefan, who had recovered from carsickness and food deprivation, got out and walked around. We opened the driver's-side door and both of us peeked into every crack and crevice we could find. Nothing! As Bill Vinci had proclaimed so loudly in the bar . . . *nada*. Finally, after five increasingly desperate minutes, Stefan found it! Some automobile interior design wizard had put a pencil-eraser-sized button into the extreme left portion of the indentation in the lower left pocket of the driver's door—usually a place for sunglasses, chewing gum wrappers, and the like. As far as I was concerned, the person who made that decision should be assigned a noncreative place in automotive hell for placing the gas-cap release down there.

One thing was piling on top of another.

After thanking Stefan, I started the pump, and then literally sprinted into the service center to get directions. I peppered the young man with questions: "Where are we? How do I get to Westborough? Why are intersections up here so impossible? What do I do now?" In my sense of frustration, humiliation, desperation, and guilt, I blurted out too much information, finally pleading for help getting back to a hotel that couldn't have been more than three miles from my disastrous decision to take Exit 22.

"Go on to the next exit," he *should have* replied, "then reverse, and go back to 495 where you got on. Take 495 to the next exit, which is 23B. You'll be on Route 9. Exit on Computer Drive . . . you can't miss it."

If he had said that, then good-bye and good luck, we would have been back in fifteen minutes and could have cut (some of) our losses. But after hearing my pleas, he told me to wait and immediately disappeared into a back office. Did I look disheveled? Was he going to get a gun to defend himself from a raving lunatic? Finally, he emerged with the most detailed set of MapQuest directions I had ever seen.

"Just follow these and you'll be home free," he said. He must have felt like the good Samaritan. But it seemed to me like I had gotten a set of clues for some sort of road rally. Hints that would result in locating some form of cleverly hidden object. Do this, do that, do twenty more things—it was three sheets long!

I walked slowly back to the car, not quite knowing what to do. First, put the gas cap back on—that was step one. (At least we wouldn't run out of fuel.) Inside the car, the anxiety was palpable. Not only were we lost, more or less, but Stefan and Manfred still had no confirmed way to even enter the competition in the morning. The rules were explicit. Without a way for Wes to monitor them both, the team couldn't take part. If we ever got back to the hotel, Stefan had to locate a soldering iron. Earlier in the evening, that might have been routine in a crowd of technical people. But it was almost midnight, and with every passing minute the solution became more iffy.

Things rotated through my mind. When we got back—if we ever did—if Stefan found someone roaming the halls with a soldering iron around midnight, if he could construct his little audio-isolation gadget in his room, and if he and Manfred ever got some sleep, would the circuit work back in the woods in the morning? And if it did work, would the guys be able to stay awake and stay sharp under the stress of the nonstop twenty-four-hour World Radiosport Team Championship event, while competing against many of the best contesters in the world?

First I had to get them to the hotel. We were back on the road now, and within five miles we exited the Mass Pike according to the printout. We were supposed to get on Route 9 and head west, back in the general direction of Worcester. The first complication arose when I pulled up to the tollbooth and realized I had never taken a ticket at the entrance point at Exit 22—in all the panic and frustration, I had simply driven right past the automatic ticket dispenser. A bored young man at the booth perked up. Aha! A special case: a drunk, or a dolt from the hinterlands. We negotiated a payment for an idiot with no ticket—he charged us $3.40. I just wanted to get out of there, but he insisted on explaining that Route 9 was very congested and we just simply had to stay on it and take it street by street until we got back to Westborough.

To say that Route 9 is crowded and stop-and-go is an understatement. Thirty minutes later, after innumerable traffic signals and what seemed to be an infinite number of side streets, stop signs, mini markets, gas stations, exotic dance clubs (all with packed parking lots), and shady adult bookstores, I was about to give up, fearing that we somehow had become lost again. No road signs indicated in any way that we were on Route 9, but we passed a vast collection of Main Streets and Maple Streets. Finally, out of the morass, a sign announced "Westborough City Limits," and shortly thereafter another sign with an arrow pointed to the exit for Computer Drive.

When I pulled the Charger into the DoubleTree parking lot, we were the only thing moving. All the spots were taken. (Any sane or at least rational person was in their room asleep.) We walked to the lobby from the "back forty acres," and Stefan and Manfred set out to seek the indispensable soldering iron. The clock at the reception desk indicated a bit past midnight. I spent a minute apologizing to Wes in the absence of the Germans, and then started up the steps to my room on the second floor. As I glanced back one last time, Stefan and Manfred were joining Wes on the elevator, and Stefan smiled and waved at me with a soldering iron in hand. *Oh, Great Radio Spirit, make it work. Make that circuit work.*

*Chapter 10*

# THE FIRST HOUR

Feelings of guilt, anger, and frustration ate away at me as I fished around for the electronic room key. All of us had agreed to meet in the lobby at 5 a.m. Would Stefan be able to cobble together his home-brew audio-isolation circuit? I set my cell phone alarm for 4:15 and punched in a zero to get an insurance call. The night-shift desk operator answered on the third ring and acted as if it were completely normal to phone at 12:30 for a four-something wake-up call. I felt too numb to be tired, if that makes any sense.

It would be impossible to get to sleep—just wasn't going to happen. After ten or fifteen minutes of being hopelessly wide awake, resigned to tossing and turning for three hours, a weird sound jolted me wide awake. The alarm on the phone sounded like a combination of *Star Wars* and Bach, somehow folded together by a synthesizer. Where was I? What was happening? Just as the fog lifted and I realized that the cell phone was the source of the interruption, the handset beside the bed started to ring.

Thirty minutes and one soul-saving hot shower later, the situation began to reset itself with more clarity. The drive to the site would take ninety minutes. So, if we left at 5:00 on the dot, Stefan, Manfred, and Wes could try out the headphone system at 6:30. If that didn't work, no good options remained—it just had to work.

At 4:45, I opened the room curtains to see first light and tiptoed out the door, trying to be perfectly quiet. It seemed surprising, although it shouldn't have been, that the lobby was full of people milling around: Teams were getting ready to leave for sites spread all the way from just below the New Hampshire border down to the state forest south and east of Rhode Island. The din of voices masked an atmosphere of overall tension. The "bad" me, the "we finally got back to the DoubleTree at midnight" me, was nervous. I was prepared to call their room if the guys weren't ready. As my mind was working through all sorts of worst-case scenarios, Wes appeared.

I didn't ask him about last night, but our eyes connected with a sort of silent understanding—it had been frustrating, to say the least. At 4:55 the elevator door slid open. Stefan and Manfred walked out, appearing surprisingly chipper. Stefan saw us and held up two small items over his head. Each one looked like a hunk of black taffy with plastic wrapping twisted around each end and with two electrical connections protruding a couple of inches.

The hotel had prepared breakfasts-to-go in paper sacks, which were arrayed in tidy rows on a table to the rear of the lobby. A huge coffee urn sat nearby. Each of us took a little tote bag and walked to the lobby, passing by other teams and drivers. Too-early "good luck" comments were mumbled as we went outside and walked to the Charger. It was 5 a.m. on the dot, and a fairly uniform early morning light was settling in. The car was covered with heavy morning dew, so I ran the windshield wipers as we left the parking lot.

Navigating the spiderweb of access roads leading to the major routes was again difficult:

- Exit the crowded DoubleTree lot.
- Access road straight ahead at the first intersection.
- At the next intersection, left over the bridge crossing Route 9.
- Left again on the other side of the bridge, now paralleling Route 9 in the opposite direction.
- Pass the turn-off into the Industrial Park entrance. Go straight ahead.

Overall, it was maddening and convoluted. But finally we came to the sign that said "495 South." So far, so good.

No one said a word. The three men were probably in a mental condition somewhere between fatigue and fear of the directionally challenged driver. No matter. Within a minute we were out in the clear, cruising comfortably at sixty miles an hour heading south.

The next hour was pleasant. All three passengers went to sleep almost immediately, while I kept the speed moderate. The hotel's black coffee was putting my brain into some semblance of operating order. The dawn gradually brightened, and when we exited 495 onto Route 58, the new day was firmly established. My tired fellow travelers continued to sleep as I snaked along Route 58 to the point where we exited onto Tremont Street. After a half mile, the partially hidden "Cranberry Road" sign appeared and I turned right and continued past the bogs. Once again, several of the big commercial sprinklers were on, and huge plumes of water were shooting out in jerky sets of pulsating arcs.

We started uphill and entered the deeply forested slopes. Several miles later the rustic headquarters building came into the early morning view. I turned to the left, focusing hard on remembered directions, so hard it almost hurt:

- Up the hill past the Hawaiians' site with the lake—no, it's a pond here—on the right.
- At the next intersection, a hard right.
- Stay to the right. Keep to the right!
- A mile later, it's starting to look familiar.
- Here's the left turn onto Fearing Pond Road beside the "bridge out" sign.

The guys were waking up now—perhaps from the stops and turns. There it was. That wonderful, weather-beaten, faded sign: "Camp Squanto." Hurrah! Turn right onto a patch of washboard asphalt, then a left turn and the straightaway. The first sign was there—right where it was supposed to be: "WRTC site 15E" for our neighbors across the road, and now the "WRTC site 15W" on the right. We made it.

Wes was the first to speak. "Nearly there. Everyone ready?"

By now the sun was beginning to clear the horizon. But as we started down the tricky forest path, the trees blocked any of its radiance into a dim remnant. After all the day and night trips, the 500 meters finally were beginning to seem less onerous. Down the dip now, across the bottom, and up the other side, where the forest transitioned into the clearing. The antenna glistened, reflecting the sun. Two tents beckoned in the distance.

Both the Bennetts were up and about. They offered one and all a cup of some serious camp coffee. It was dark and gritty, with flecks of beans that gave their all for the cause. Everyone but Manfred, who didn't drink the stuff, took a cup. Then the Germans went directly to the operating tent to try the new audio circuit.

A few minutes later they called for Wes and me to come over. The test was at hand. Manfred and Stefan first connected without Wes; each could hear his own radio but couldn't hear the other's. So far, so good. The moment of truth would be with Wes connected in the middle. The atmosphere was tense. Wes signaled with a thumbs-up that he could hear the signals from both receivers: Manfred's radio was in his left ear and Stefan's was in his right. The signal levels to both radios dropped a very small amount when Wes connected his earphones. But that was normal, and they could compensate by increasing their audio level a bit. It was okay—Stefan's new audio-isolator circuit worked!

Now it was the calm before the storm. Not a meteorological version, but the collision of thousands of invisible strands of electromagnetic energy that soon would be flying from fifty-nine teams to nearly 10,000 individual ham radio enthusiasts worldwide.

Narrow rays of sunlight peeked through the pine trees, confirming the fair weather that had been forecasted—at least for Saturday and early Sunday. One storm system had passed earlier in the week, and a new weather front was developing—one that could hit eastern Massachusetts as early as Sunday night or Monday. The forecast was welcome, even if it meant that daytime temperatures would be in the high 80s, making the heat in the tent oppressive. Speckles of sunlight started to dapple the tent as the team walked back to the campsite to open the hotel breakfast bags. Judy Attaya-Harris's Eurocentric shopping trip the previous day had been centered mainly on food for lunch and dinner, but there was plenty of orange juice, which went nicely with the hotel's breakfast tacos.

The station was finally ready to go. Wes clearly was elated. Outwardly, he had seemed the most concerned of all. Of course, he had done nothing wrong. But no doubt he didn't want to be forced to tell his team, "Sorry, no go." Yet the rules were explicit: He had to be able to monitor both operators, at least one at a time. Perhaps they could have developed some sort of last-minute, slapdash method to allow him to do that, if Stefan's clever new isolation creation hadn't worked. But it did. And at 7 a.m., with one hour to go, a fly on the wall would see six men laughing and relaxing while eating breakfast tacos at a campsite deep in the woods.

It was more-or-less common knowledge with the cognoscenti, or at least best-case forecasting, that as many as 5,000 contacts might be achieved by the top teams. Of course, that assumed optimum conditions, along with weather that would hold up with no thunder, static crashes, or atmospheric disturbances. Straightforward math yielded an average hourly contact rate of slightly more than one hundred per hour for each operator—nearly two contacts every minute for the entire twenty-four hours. The Germans hadn't discussed their plan with Wes or me; however, in a competition like this, nearly every team would be thinking along the following lines:

- For the highest rates of contacts, teams would call CQ. At times the callers would come single file. That was the ideal case, but they often would come in multiples. Large numbers were called a "pile-up."

- When the CQ rates decreased or conditions simply were poor, it would be important (and at times more fun) to tune your receiver and search for other stations to call. This was especially productive when looking for more unusual or rarer stations that always seemed to have their own strings of callers. This was called "search and pounce," since you were on the prowl. Find 'em and pounce!

- Emphasize Morse code over voice (phone) contacts on the assumption that low-power (100 watts) would "get through" on Morse better in marginal conditions.

- If a team found a clear frequency, and if it stayed clear with their low-power and simple antennas, then it could be possible for voice contact rates to reach 200 per hour for a single operator. In that case, to "strike gold" for an hour or two would be a key advantage for any team that could make it happen. But it would be likely that a stronger station, with more power or better antennas, could barge in directly or indirectly—take over the radio frequency. In that case, the operator would have to be nimble, not stubborn, and try to find a sheltered "harbor" frequency quickly because their "liquid gold" stream of callers could be shut off fast.

- A new software program allowed ham radio receivers, which had been set up at stations around the world, to detect Morse code signals automatically. The dits and dahs, no matter what the speed, now were converted to normal text and "spotted" on the Internet. A team's Morse signal would be relayed widely. Amateur radio operators worldwide could, with one click of a mouse, send their radios immediately to that very "spot," as it's called—to that team's exact frequency. This meant a huge following for the WRTC competitors on Morse.

- In the daylight hours at the start of the contest (8 a.m. local time), and continuing through late afternoon, teams would concentrate on the "daylight bands," those radio frequencies that reflected off the outer fringes of the atmosphere. These very thin outer parts of the ionosphere, supercharged from the emissions of solar energy, became a reflective mirror for the transmitted waves. Three main "daylight" bands were part of the WRTC, and the teams had to pick the best two of them at any time, and hit them as hard as possible for as long as possible.

The contest score, in overview, would be a product of the number of contacts, multiplied by the number of points per contact (more points for a greater distance), multiplied by the "multipliers," which were different countries as well as headquarters stations from different national radio societies.

(Score) = (# contacts) X (points per contact) X (# multipliers)

Therefore, it would be important to make as many far-reaching contacts as you could, but also with as many different countries as possible, on as many of the five WRTC contest bands as possible. These multipliers would count separately for each of the different frequency bands in the championship: the three daylight bands, and the two nighttime bands. Two of the three daylight bands were usually predictable—they would be open, and most likely they would be jam-packed with action. Both of the nighttime bands were fairly predictable.

Several major variables and uncertainties would be in play:

- The dawn and dusk periods could be most intriguing, in that they combine "daylight" and "dark" solar conditions. These exotic and unusual transitions could occur along the so-called "gray line" where daytime and nighttime atmospheric conditions co-mingled temporarily. Experienced operators could find hard-to-reach locations during these periods of a contest.

- In the summer, the energy from the sun is intense and lasts quite a long time in the northern latitudes, including North America and Europe. As a result, the ionosphere remains energized and at least one of the "daylight bands" could remain open until late, sometimes nearly all night. The competitors must keep an eye out for this occurrence. In these situations, the two main "money" bands overnight would be the prime-time nighttime band as well as the daytime band nearest in frequency to it.

- Another big question would be the daylight band that is the highest in frequency, the one farthest away, in terms of frequencies and radio physics, from both of the nighttime bands—the most extreme "daylight band," so to speak. Depending on the eleven-year sunspot cycle of activity, that group of frequencies could be a total dud and produce nothing but white noise in the headphones all weekend. Or it could burst alive and go completely nuts, remaining open to vast stretches of the globe with signals so strong they boggle the mind. Occasionally, it would lurk in apparent slumber, but then open its sneaky eyes just a peep to allow for marginal conditions—but only for minutes at a time. The smart money would be for the teams to check on that band (28 MHz) often in order to be sure they didn't miss an opening.

A handful of the teams might go for more voice, or "phone," operation than the Morse code option. (Several operators were notably stronger on phone.) They would be banking on the potentially higher rate of contacts—in addition, there would be a major prize for the team with the highest number of voice contacts. Yet every knowledgeable radiosport observer predicted that a "Morse-centric" strategy was the right way to go, even with a few selected boffo runs of voice during optimum conditions.

---

At 7:30, Manfred and Stefan finished their final pre-contest food ("meal" is too elegant a description) and slowly walked the hundred feet or so to the WRTC tent. Those leisurely steps might be their last ones outside for twenty-four hours, except, of course, for nature breaks. So, they took deep breaths and tried to enjoy the beauty of the trees lining the edge of the clearing. The sun had now risen above the forest and reflected off the new aluminum elements of their antenna, which dominated the site. Sounds of the forest provided the music of the morning: a bird chirping, a soft breeze through the pines, and nervous squirrels chattering somewhere. For the following day (and night), the contestants would hear thousands of snippets of human voices in a hundred accents and dialects of English. They would hear Morse code sent mainly by computer but also by a minority of purists and old-timers who produced their dits and dahs by hand with an assemblage of mechanical gadgets.

They pulled the tent flaps aside and sat down in front of their computers. Within a minute the two large monitor screens came alive. High-resolution color displays showed a sea of information that would overwhelm most people. One window included an electronic logbook showing where every individual contact would go: the station's call sign plus the country. In addition, a world map showed the "gray line" between day and night, as well as a separate summary of the number of contacts and multipliers on each of the five frequency bands. A small fan blew a gentle stream of still-cool air.

At exactly fifteen minutes before 8:00, Wes, as referee, reached into a knapsack and drew out the magic envelope. This was the moment to find out what call sign the team had been assigned. With a friendly flair for the dramatic, he slipped it open and produced a stiff card.

W1P

The letters were written in a large, bold font. Each team's call sign would begin with a *W, K,* or *N,* the "prefix" signifying a US station. The number in the middle indicated "US district number 1," which covers all of New England: Connecticut, Rhode Island, Massachusetts, Vermont, New Hampshire, and Maine. The single letter at the end had been selected by the WRTC 2014 organizers.

Stefan and Manfred's call sign was W1P. It sounded easy to me: Whiskey One Papa. Since English wasn't their native language, the two men looked at me quizzically. They wanted to know the best way to say the word *Papa.* (I assumed most everyone knew how to say *Whiskey.*) Stefan asked, "Is it Whiskey One Pa-*Pa?*" The emphasis was on the last syllable. Manfred seemed to prefer "Whiskey One *Pa*-Pa," with the first syllable hit harder. I recommended a bit of a hybrid, with *"pa-pa"* having just a bit more emphasis on the first syllable but nearly the same, and crisp and clear. They practiced it aloud over and over, and when they were satisfied they loaded their own voice messages into their individual computers. This would save their energy, improve their stamina, and reduce ambient noise in the tent, allowing each man to concentrate on his own receiver more effectively.

This little burst of activity took up several of the fifteen minutes of "quiet time," when they weren't allowed to listen to any of their radios prior to the start of the contest. They sat there, ready to go, primed with adrenaline, but now in a bit of contemplative silence as the tent became more and more infused with the morning light. I checked the clock on the desk: 7:55, 7:56, 7:57—only three minutes to go now.

The atomic-clock digits blipped from 7:59 to 8:00. Wes gave a bit of a theatrical hand signal. Manfred and Stefan put on their headphones and tuned their receivers in search of a clear frequency they could use to send out the first calls.

At that point, the WRTC transitioned from interaction with the team to separation. The "in-and-out flap" of the tent now gently lay in place, and any view from the outside passed through mesh net openings on all four sides. For the next twenty-four hours, according to the rules, I was only an observer.

From this moment on, none of the actual team members inside the tent were to know how they were doing in comparison to the other teams. According to the guidance provided in the referee meetings, specific ranking spots were not to be shared with the competitors, even if somehow the referees themselves were informed while outside the tent on a break! None of us outside the tent had any way to monitor the team's progress other than the Internet scoreboard. Fortunately, Mike Bennett's cell phone had a strong connection with his carrier's base station—wherever that was—so we could check the results at all times. For the first hour of the contest, not a single, solitary word came from the tent. Only faint metallic clicks of the Morse code paddles—those precise mechanisms of metal and plastic—could be heard, though not any Morse itself, and only if you stood close to the tent flap door and listened intently. Lots of things were going on, but there was very little physical movement, and no sound outside the earphones.

The W1P team made 232 contacts in the first hour—all on Morse code—142 on the 14-MHz band, and 90 on the 21-MHz band. The initial one-hour results were good but not great. Stefan and Manfred were in the number-13 position.

Both of the closest neighboring teams in the Myles Standish State Forest, located on the road out to Camp Squanto, got off to a better start. The Americans, Dan Craig and Chris Hurlbut, who were favored by many to win, were only 500 meters away to the east and had opened up their envelope to find K1A. That was a primo call sign anyone would like to have: easy to say on voice and snappy on Morse. They began very strong in terms of contacts, with the highest number of any team, at 253. Yet, since these included only 52 multipliers, they sat in the number-6 position.

Across the road was one of the four sponsored teams, Latitude 13. (The name was probably coined from the northern coordinate of Barbados, where Tom Georgens operated—and dominated—and where he had racked up some of the most impressive radiosporting scores of the past decade.) In the first hour, he and Dave Mueller made 214 different contacts, which included sixty multipliers—a solid number. They were looking good in the number-8 position with K1S, the call sign assigned to them.

▷⟶ℓℓℓ⟶▮▮

Sandy Räker and Irina Stieber had a problem-free setup at their site near the old state hospital. Tim Duffy, who in many ways was "Mr. Radiosport" in the United States, made sure the complex configuration of equipment went together smoothly. In their straightforward on-air trial Friday afternoon, the women had been besieged by large numbers of American stations calling them, and the reports of their signal strength and quality were encouraging. Räker had a US call sign, N0QQ, and she went on the air in that way. Her bright smile appeared to translate to her radio signal, based on the flood of responses. Stieber used her German call sign, DL8DYL, or "DL8 Deutsch Young Lady" operating portable in the United States, and both were pleased and a bit surprised at how well their stations performed. Their site managers were "awesome" (Räker's own word), taking photographs of the site in different sunlight angles. When the team arrived early Saturday morning, a cooler in the tent was already there, nicely stocked with water and soda pop. The women had requested no food at all; they brought their own "stuff," a scant collection of M&M's, Oreo cookies, peanuts, and carrots. Duffy's steady support and quiet assurances, along with referee Rusty Epps's exuberant sense of humor and experience, were reassuring. The women had an A-team in addition to a supportive site group.

Räker and Stieber sat and watched as Epps opened the envelope. Wires and cables ran everywhere, connecting the two computers, two radios, the magic little scoreboard score-collector gizmo, and three pairs of earphones. At the top of the standard-issue tent, a crinkly aluminum reflector was positioned to deflect the solar heat, which was noticeable even now, at 7:45 in the morning. With his usual élan, Epps held up a card with the team's assigned call sign:

N1A

"November One Alpha. That's really good," he said, with an encouraging manner, like everyone's favorite uncle.

---

The official International Civil Aviation Organization (ICAO) actually uses "Alfa" as their phonetic for the letter *A*; however, most people think of "Alpha," and I have used the latter term here.

---

The pronunciation of "November" and "Alpha" was standard and easily transferable between German and English. The women immediately set about programming it into their respective computer Morse message memories and voicing it into the digital voice processors as well. With ten minutes remaining until the contest would begin, they must have felt ready to go, yet also nervously bouncing on pins and needles. Based on Räker's radiosporting experiences to date, which had earned her the coveted qualifying spot, her voice capability was a strong point. Her past experience as a high school exchange student in Minnesota had helped form a pleasant accent in English, definitely American rather than tinged with the British English of many who study in Europe. Räker had worked on improving her code skills during the lead-up to the WRTC.

Stieber, who came from the exact opposite radiosporting background (with her world-class skill in very, very fast code), had less experience with the English language and spoke with a fairly distinct German edge. She had worked to improve both her pronunciation and her overall comfort and ability with phone contesting.

From their own research, from conversations with the two German men's teams, and from strategy discussions with Duffy and Epps, Räker and Stieber knew well that Morse code should be their focus. Overall, they felt they could operate in either mode as the conditions demanded.

Epps flashed the "go" sign at exactly 8:00 local time. Räker and Stieber started on two of the daylight bands and racked up 105 and 57 contacts, respectively—all on Morse code—for a total of 162, along with fifty-eight multipliers. After an hour, they were fortieth overall, out of fifty-nine. The station seemed to be working well. So far, so good.

The number-1 team in that initial hour was the Slovakians, who were operating ten miles away at Freetown Breakneck Hill, site 14A. The remainder of the initial lead group, clearly off to terrific starts, were an American-Canadian team in the second position, as well as an Anglo-Irish team, two Asiatic Russians, and two Bulgarians to round out the top five. The strategy was consistent: Every team was focusing on the top two daylight frequency bands.

These early leaders made a total of 1,191 contacts altogether, with only forty-five on voice. The race was on, and all of the racers appeared to have a similar plan so far.

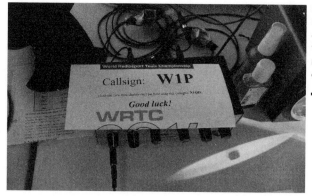

Suspense broken . . . the card with the W1P call sign. Photograph: Jim George (N3BB).

L to R: Manfred Wolf (DJ5MW) and Stefan von Baltz (DL1IAO) operating in daylight. Photograph: Jim George (N3BB).

Irina Stieber (DL8DYL) and Sandy Räker (DL1QQ) at their operating position. Photograph: Michael Hoeding (DL6MHW).

# Chapter 11

# AND THEY'RE OFF!

Not every team started primarily on Morse code. George DeMontrond and John Crovelli had made almost half of their 141 contacts the first hour on phone. That reflected on DeMontrond's specialty. The bearded Crovelli, who could have passed for a beach bum or a Silicon Valley tech maven, had decided to work hard on the team, and operate in a way that was complementary to DeMontrond's focus on voice operation. Although he called CQ and ran the pileups on Morse code, part of the time he listened for needed multipliers as well.

DeMontrond was very good on phone: "Inviting and engaging. People want to call him," according to his teammate. His style, however, was "Loud—he's a screamer," and they had worked on setting the transmitter gain and the microphone compression settings for both "oomph" and clarity. The team had to deal with distorted audio at one point, but they were able to settle down with a proper-sounding voice signal. Unlike the Germans, they had no trouble providing audio: They used an American-made selection box with "left, right, and stereo" options, plus a twenty-foot cord for the Ukrainian referee's headphones.

Crovelli had years and years of both voice and Morse contesting at the highest levels. A human atavism of sorts, he preferred to send code using his own fingers to make the longer dahs and the shorter dits with a paddle connected to an electronic circuit that keyed the radio. This was in contrast to most teams, who simply typed on the keyboard, used prerecorded Morse messages, or did both. Crovelli has estimated that overall he sends at least 80 percent of all his transmitted code by using this "sort-of-by-hand" method. For the present competition, he set the speed to thirty-six words a minute and then "never changed it at all." He was a throwback to an earlier era, typing with two fingers on one hand. "It's second nature to me," he said. "I just like sending code."

⊳〰〰〰┉

The Lithuanians had been shaken by the dead PC on Friday, but the borrowed Dell seemed to work. Back at the site on Saturday morning, in the early morning light, the computers (both of them) and radios all came up to speed without a hitch. Everything appeared to be set—quite a relief after what they had been through. Fifteen minutes before the starting time, Gediminas Lucinskas (everyone called him Gedas) and Mindis Jukna put everything down and watched with excitement as John Laney, their referee from Georgia, opened the envelope.

WIA

"Whiskey One Alpha," Laney said in his southern accent. The call sign was clean-cut to say on voice and snappy to send on Morse—another good development. Everything was coming together at last. When Laney gave them the go-ahead at 8:00, the Lithuanians started. The contacts came fast and furious on Morse, as they had planned. Soon there was a steady stream of callers for both operators on their targeted daylight bands.

Within a few minutes, however, Gedas's borrowed Dell started "lugging." For some reason it would take a second or two to enter the data into the electronic logbook. A snippet of time like this might seem like a trivial annoyance. But when the entire two-way contest exchange at Morse code speeds of forty words a minute takes only five seconds, it's an eternity. Another station—or more likely a small howling pack of stations—would be calling and Gedas would have to sit there, bursting with frustration, before the computer gulped down the previous chunk of data and burped out the call sign of the new station he had just decoded in his head and typed into the keyboard. The timing problem completely botched up the rhythm. In many cases, the callers would become impatient and start to resend their call signs. All over again, "It's me, it's me," the horde would scream in desperate (and disparate) tones, speeds, and Morse personalities.

When his computer finally swallowed that last chunk of information, and often exactly when the "pile-up" called a second time, his station would send the delayed information. The confusion was bad enough when one or two seconds had elapsed. Yet the problem grew to five, then ten seconds. These delays made it impossible to use the computer to transmit anything, even when it finally put the callers into the electronic records. Looking for something, anything, that would work, Gedas started using his Morse "paddle" to send everything by hand. He was frantically trying to sort out the pack of callers and then send the information manually. But the computer got so slow in adding each new contact to the database that finally it was completely impossible. Gedas momentarily started logging on a piece of paper before realizing the situation was hopeless.

John Laney watched and listened to all of this. It was extremely frustrating to see such a fine radio operator get dragged down by the failing computer. Gedas and a Lithuanian friend conferred in a blizzard of their native language. As a possible solution, they sought and received permission to disconnect the scoreboard computer, thinking that it was somehow contributing to the delays, and then tried to reload the contest software on the Dell.

Finally, the screen showed that there were nearly 900 "issues." It looked hopeless. Gedas took off his headphones and finally shrugged in despair. Fortunately his teammate Jukna's systems still were all "go," and he was notching callers one after another. At least one of the two stations was okay. But the team would soon be left in the electromagnetic dust if they couldn't come up with a solution.

⊳‑⅊ℓℓ‑⊪

The American team of Kevin Stockton and Steve London at the W1Z site ended the first hour at the number-19 position. They had no way of knowing that, since specific ranking information from the scoreboard wasn't relayed to the teams. Their equipment was working fine—no problems at all—and each operator, focusing on one of the two daylight bands, had had 124 and 102 contacts, respectively.

They had been the penultimate team to draw. London hadn't worried about the location. His main concern was not having much interference from nearby competitor teams. There was only one other team at their site, a Bulgarian group, and they were nearly three-fourths of a mile away. The site appeared to be on completely level ground in the direction of Europe, but with a bit of a bowl to other locations. Yet London was satisfied. Twenty-four hours would be a long, grueling time, but he was ready to dig in and work hard. He was known as a relentless operator with almost superhuman grit and stamina.

⊳‑⅊ℓℓ‑⊪

The defending champions from the WRTC 2010 in Moscow didn't like much at all about their location near the taxiway at Mansfield Airport. The site selection committee had located the flat area in the summer of 2012 and tested the signals into Europe using receiver reports in a trial run that same summer. Vlad Aksenov and Alexey Mikhailov lived in the northwestern part of Russia, and both were experienced elite operators (with the gold medal from 2010 as proof). In addition, both were strong technically. Aksenov, the qualifier, was a 47-year-old radio-frequency (RF) engineer and director of a commercial amateur radio company. His partner, Mikhailov, at only 35, was already the founder and CEO of a telecommunications company.

Both men usually operated at RU1A, one of the world's largest—and in some ways most secretive—contest club stations. At least one well-known American contester had visited the complex as Mikhailov's guest. He was picked up at his hotel in a new 7-series BMW. Alexey (ham radio operators are always on a first-name basis when together) sat in the front passenger seat, and a most definitely all-business driver wore a pistol and said nothing. On the drive to the station, the car passed through "very Russian-looking" villages and towns with Soviet-era apartments as well as many farms on the way to the city of Luca, located two hours south of St. Petersburg. "It is important, do not speak at all" were the stern instructions to the American as the luxury car pulled onto a dirt road five miles to the west of Luca and approached a military checkpoint. Alexey and the driver presented some forms of identification to the guards and spoke briefly.

The car was waved through to "twenty-five or thirty acres of land, including two or three houses" on what seemed to be a privately owned civilian compound that somehow existed on the fringe of a military complex. At least fifteen huge towers, apparently military in nature and of square construction (not the normal triangular shape), between 150 and 200 feet tall, loomed up through the trees surrounding the houses. Most of these massive amateur radio antennas were rotatable, but several were positioned in a side-by-side fixed manner to maximize the signal strength to the United States.

Not one, but *two* large eighteen-wheeler-style cargo trailers supplied electrical power, apparently through industrial or military generators. The American described the operating room itself as "something out of a James Bond movie," with a large number of state-of-the-art amateur radio transceivers, world maps, time zone information, and the instantaneous day-night "gray line" information. Alexey was "a very gracious host in all respects," according to the American visitor, who said that he could hear what appeared to be rumbling, deep-throated sounds of tanks maneuvering off in the distance. When asked about amplifiers, which would boost the power from 100 watts to higher levels, the American said there were none in the operating room and that he neither asked to see, nor was shown, any other equipment. Based on the amazing signal levels from RU1A in contests, it's safe to say that the station was "fully equipped."

As might be expected from such experienced competitors, the Russian team opened up in the WRTC solely on Morse code and focused entirely on the two primary daylight bands. They knocked out the sixth highest number of contacts on one band (14 MHz), at 143, and together with sixty-nine on the 21-MHz band they ended the first hour in the number-14 overall position.

The Estonian team members, winners of the silver medal in Moscow, definitely weren't satisfied with site 5B, one of the northernmost locations, called Twin Valley East. On Friday afternoon, as part of their equipment setup, they'd had a problem with one of their filters (these kept signals on one band from interfering with other bands) but got that problem fixed. More of a concern to team leader Tonno Vahk were the disparities he found when he compared his signal with that of other WRTC teams, when two teams were on the air side-by-side. Most of the signal checks were equal; however, a couple of the reports (on voice mode) indicated that the Estonians' were down, or not as strong, by a significant amount. Those reports, along with the very slight rise in the topography in the direction toward Europe, left a question in Vahk's mind.

The team's 178 contacts in the first hour were below most of the "big names" and landed them in twenty-fourth position. Vahk later reported that they called CQ without answers—"a CQ fest," in his words—during the early hours. Vahk felt they weren't "getting out" very well, in radio-operator speak. Vahk started the contest on voice on the 21-MHz band, and had no contacts at all for the first four or five minutes! Following that, he spent more time tuning around for stations to call before he moved over to code. His teammate also operated both Morse and voice the first hour on 14 MHz. Unlike most of the experienced operators, they made 28 percent of their contacts on voice. Overall, the start of the contest wasn't good for them.

Ken Low and Scott Redd had arrived early at their Fort Devens site. They were relieved when all the equipment powered up without problems. With the new grounding system, the troublesome noise hash on one of the wire dipoles was gone for good. A decent, albeit short, night's sleep soothed the fear of being overpowered by the strong neighboring signals from the Brazilians, located only 500 meters away. This would be the first WRTC for both men.

At the "go" signal from their Serbian referee, Low and Redd started on Morse code on both primary daylight bands. They racked up a strong tally of 254 first-hour contacts, with 155 on one band and ninety-nine on the other. Things were moving fast. They were in eighth place and solidly positioned, but of course they didn't know that. The initial strong start fit exactly into their strategy: Emphasize Morse code because the worldwide software-decoder system would pick up their call sign and frequency and blast the information on the Internet. They would stay on a particular frequency band as long as callers kept coming hot-and-heavy, and would change only when they felt the contact rate was dropping. In addition, they would spend some time tuning and searching for stations in new (and hopefully rare) places, but primarily would expect most of the multipliers to find them. Redd in particular excelled at sitting on one frequency and making rapid contacts one after another on Morse. He had a strong "keep the rate up" attitude and would be their anchorman (pun intended for the retired navy man). They were off to a fine start.

⊳⌐ℓℓℓ⌐╫⊪

The retired broadcast engineer, John Barcroft, and his former horn-playing classical orchestra member—and now broadcast TV engineer—David Hodge were located near the New Hampshire border at one of the two Twin Valley sites. They ended the first hour with a solid 215 contacts, nearly all on Morse, and were only one spot behind their American counterparts, Kevin Stockton and Steve London, located a few miles away with the call sign W1Z. It was a good start for Barcroft, who had worked so hard and in such an organized way to qualify.

⊳⌐ℓℓℓ⌐╫⊪

With fifty-nine teams, of course not everyone could be on top, even if each and every team in the field had been the top dog in their region of the world. Don Kondou and Hajime Hazuki would have been considered long shots, since topflight radiosporting in Japan was more difficult due to the dense housing and high cost of land. In spite of this, there were a few contest clubs in the country, with a handful of active individual stations near Tokyo and Osaka, as well as on the northern island of Hokkaido near Sendai. Yet, all in all, the concentration of "radio active" enthusiasts in Japan was less than in hotbeds like Central and Eastern Europe and the United States.

Kondou had overcome terrific obstacles to first reengage in the hobby and then qualify for the world championship. He was the sole user of a Mac-based competition logging program named SkookumLogger; in fact, the software developer himself, Bill Myers, K1GQ, had volunteered as the team's driver and on-site software and computer technical support. Kondou had become one of three key beta-testers of SkookumLogger since its introduction in early 2009, and this would be a high-visibility introduction into the normally PC-centric world of radiosporting. The word *Skookum* means "impressive, big, strong, and durable" in the native Chinook language spoken in the Pacific Northwest of the United States and Canada. This would be an outstanding test for the program, since Kondou and Hazuki both used Mac computers and ran their software on the Apple operating system.

Hazuki had planned their operating strategy for the WRTC, even though he lacked Kondou's on-air operating experience. They would look more for other stations to call, including elusive multipliers, by tuning for them, *searching* for stations, then *pouncing* on them and calling. Of course, it was necessary for every team to have some S&P, as it's called in shorthand. Yet the best scores would be mostly high-speed runs, when people were calling one after another, along with some judicious tuning around.

Their station setup was in "central middle" Wompatuck State Park, on Boston's south shore near Hingham, and was the last "big one" in terms of potential WRTC operating locations found. In fact, the site finder team had come across it in the summer of 2013, when they were getting short on time. The area stood out visually. "I poked and prodded terrain maps and satellite maps feeling like a Martian looking for an open spot to land on Earth" was how Rich Assarabowski, K1CC, described his method. "A big patch of green (4,000 acres!) just fourteen miles southeast of downtown Boston caught my eye. It was a state park, a former World War II navy ammunition depot with an interesting past." I'll bet! "Small open areas scattered in the dense trees turned out to be bunkers, connected by overgrown railroad tracks and power line poles rotting in the woods." His first estimate at operating sites, each of which had to be accessible and separated by at least 500 meters, resulted in fifteen good candidates.

It was one thing to see potential areas on a terrain map, but yet another to reach them and do an on-ground assessment. All potential locations were behind closed gates, and the only transportation option was bicycles. On one long and exhausting day, Assarabowski and Greg Cronin, W1KM, bicycled from one site to another and took extensive notes. Many additional trips ensued, resulting in a final list of eight sites in all. In the end, after dealing with buried rubble and old bunker foundations in the ground—problems that might prevent tower and antenna placements—their final portfolio of WRTC sites added seven "for sure" and one spare, located near the coast.

Kondou and Hazuki set up their extensive gear without problems, including two Mac computers with the SkookumLogger program. On Saturday morning, they were all set to go. In the first hour their strength (and weakness) became apparent as they logged 120 total contacts: ninety on one daylight band, and thirty on the other. They made nearly all of them on Morse, with only seven of the total on voice. Apparently, they were tuning for stations and doing "search and pounce" more while "running" less, since they had a relatively high number of multipliers. They were competing, and in the game. Yet, with less than half the volume of contacts, they were running far behind the leaders. After the first hour, they were in the number-58 position.

▷─◿◍◍◟╫◍

The Youth Team, at site 15G deep in the Myles Standish State Forest at South Charge Pond, fared a bit better. They finished the first hour in the number-45 spot—off and running with 177 contacts: 111 on one daylight band, and sixty-six on another. All but two of the 111 were on Morse code, so that strategy was consistent with the leaders; however, operation on the other band was split forty-three on Morse code and twenty-three on voice. Maybe they were giving voice an early trial to see how productive it was. Or maybe they lacked the experience to know they should focus on code at that time.

▷─◿◍◍◟╫◍

In the parlance of thoroughbred horse racing, "They're off!" The WRTC 2014 was underway. Back at headquarters, the organizing committee brain trust anxiously watched the scoreboard for activity and waited for the referees in the more remote, cell-phone-limited areas to call in with initial reports. In addition, they followed Internet reports closely. The automated Morse code "Skimmer" receivers worldwide were sending in reports, including the relative signal strength of the individual signals from the 100-watt transmitters and standard-issue antennas. Reports from Europe, Africa, and North and South America were nearly instantaneous, and only a tiny handful of the teams weren't shown as active. Were there problems? As it turned out, a few teams had started on their microphones with voice, and one Italian team remained on the phone mode the entire first hour, making each one of their 140 contacts in that way. That team had decided to go for the special award for the most voice contacts, rather than try to win outright. When headquarters telephoned the Italians' site to ask their Russian referee if they were having technical problems, they learned of the strategy and could relax.

Each of the teams was on the air. Overall, it was a grand start to the twenty-four-hour marathon of the World Radiosport Team Championship.

L to R: Gedas Lucinskas (LY9A) and Mindis Jukna (LY4L) with referee John Laney, K4BAI. Photograph: Bob Wilson (N6TV).

L to R: Defending champions from WRTC 2010 Moscow: Vlad Aksenov (RW1A) and Alexey Mikhailov (RA1A) with referee Denis Pochuev (K7GK). Photograph: Bob Wilson (N6TV).

L to R: "Don" Kondou (JH5GHM) and Hajime Hazuki (JA1OJE) at the opening ceremony with the Japan banner. Photograph: Bob Wilson (N6TV).

# Chapter 12

# SATURDAY MORNING

The competition portion of the big event was now in full swing. Yet what a strange "swing" it would appear to most observers. A cone of silence enveloped many of the fifty-nine team sites. To a casual visitor, virtually nothing appeared to be happening: Three people sat at tables positioned side-by-side with a fan quietly whirring, and in most of the tents no one uttered a single word. The referee was off to one side. Or perhaps he was sitting in the center with his thin audio cable that allowed him to hear either or both stations, whether on Morse code or voice. Where exactly was the excitement with all of this?

As mentioned earlier, to many competitors radio contesting is like a video game without the video. Think of it as an audio game; the *sounds* in this game include an astounding range of dialects. Sandy Räker and Irina Stieber at the NıA station would have to navigate audibly through the unique expressions of the boroughs of New York City, with "pahk the cah" from Boston, the nasal twangs and broad *i*'s of eastern Kentucky and the West Virginia mountains, and on to the low-country verbal syrup of coastal Carolina and Georgia. They would have to be quick to recognize the "*Meesh*-a-gens" and the "Wis-*kon*-sins," often spoken quickly with lots of interference from other signals, or atmospheric noise from a distant storm.

Räker's year in "Min-i-*soh*-tah" would have helped her decipher many of the American accents. She was already familiar with various intonations—the excited accentuated manner in which Italians speak English, the smooth French sound, the Asian difficulties with the "*L*" enunciation, and the Germanic tendency to turn an *ess* into *zess*. Of course, she had had plenty of experience with the European tongues. But still, she had to be ready for the flood of different inflections when she was "running" huge packs of excited Americans after the propagation gods turned off the openings to Europe.

Their equipment worked flawlessly, thanks to the technical support of Tim Duffy. And Rusty Epps's soothing presence as "Uncle Referee" was reassuring, in addition to making sure everything was done "according to Hoyle" at all times—after all, Epps was a Life Master at bridge and was used to high-level competitions with periods of silence and sophisticated intellectual components.

One hour into the contest, the German women, located near the old state hospital, were in the game. The station was working flawlessly, and they had spent the initial hour entirely on Morse. They probably sensed that they were doing well, but they had no way of knowing, as the sun rose higher over the site and morning now transitioned into midday. That morning, the signal path to Europe had been terrific, and the women improved their position, moving to number 30, then 29, and all the way up to number 22 by noon, before sliding to number 24 five hours into the contest. By that time they had made 828 total contacts. Nearly all of these were on the two prominent daylight bands, but they also had tried the "wild card" band of 28 MHz and had teased out twenty-nine contacts there. This band, highest in frequency of all those available to the competitors, could be the best—with incredibly clear and strong signals raining down from the ionosphere for hours on end. Or it could completely frustrate radio people worldwide and not open up at all.

Simulations of sunspot activity and other atmospheric conditions had been spotty (insert pun here) at best, and they indicated a possibility of some intermediate Europe-to-America openings. But there was no way for the teams to know in advance. A sunlight path was essential, so the band openings would occur when the sun was up in New England as well as in Europe. In the lengthy summer days in the northern latitudes—New England and a large part of Europe—this left a large number of hours possible. But as soon as the sun began to set in Europe, the chances would go down—and fast. Many teams believed that the best signal path would be when the sun's energy was strong at both ends, which would correlate to late morning in New England and late afternoon in Europe. The highest of the daylight bands could be critically important, since a team couldn't miss what might be a gusher of loud, easy contacts if it did burst open. In addition, since the various multipliers counted separately on every band, it was crucial not to miss them here. At 1 p.m. the WRTC teams were all over the map in terms of success up there on the 28-MHz band, with numbers ranging from a high of seventy-one contacts to a low of eight. A team's ability to catch every opening possible would be a major positive—or, conversely, a heartbreaker for some—as the contest progressed.

⊳-〰-▮▯

To the extreme southeast, in the woods way out Fearing Pond Road and then Cuttersfield Road on the way to Camp Squanto, "my team" pounded away. (Becoming "embedded," as I was, had led me to a feeling of sameness, based on shared experiences.) Nearby, two other American teams, as well as duos from Croatia, Slovakia, China, and Germany, paid attention to the rising sun as the temperature in their tents rose rapidly. And they continued to focus completely on the two primary daylight bands while keeping an eye—or, better said, an ear or two—on the persnickety 28-MHz frequencies.

For Manfred Wolf and Stefan von Baltz, more than fifty collective years of radiosporting was driving their focus. Their gear was working fine, and no problematic noise or external interference had appeared. Yet their initial ranking would have disappointed them, had they known it. Their 13th position after the first hour improved to number 8 after the second. The site managers and I had that information based on the Internet scoreboard, thanks to Michael Bennett's phone. At that point, the team had made no contacts on voice, spending all their time on code, and their 432 contacts were split nearly equally between the 14-MHz and 21-MHz bands, with only six made on 28 MHz. Was the "wild card" band open? Since there were only two multipliers out of the six, it was likely that Manfred and Stefan were reaching only local stations and not Europe.

In the third hour, between 10:00 and 11:00, they fell all the way back to number 17, making only 125 additional contacts—a sharp drop-off from the team's prior average of over 215 per hour. They must have been trying other methods, because fifty of the 125 were on voice. Was it difficult to maintain the rate on Morse, or was switching to voice a mistake? As a reference, the team in front at this point, the Dan Craig–Chris Hurlbut duo at K1A, only 500 meters east of them, stayed almost entirely on code through the first three hours and had 719 contacts compared with the Germans' 557, with only nine of those on voice. The Americans' third hour was a stellar 208, which might have indicated that it was an error for the Germans to put an emphasis on voice so soon.

Manfred and Stefan remained in the seventeenth position after four hours, with 767 contacts. Only seventeen of these were on voice. They had returned to code and were back above 200 per hour, but had not gained on the pack. However, by 1 p.m., five hours were in the logbook and they'd jumped all the way from 17 to number 4! Apparently, they found a "sweet spot" on 21 MHz, making a number of voice contacts that included several new countries, since their total multiplier count now was 163. They were in the top five!

Dan Craig and Chris Hurlbut next door at K1A also had discovered the 21 MHz opening on voice. They had made an incredible ninety-three additional voice contacts out of 210 that hour to not only remain on top, but break out with a lead of more than 100,000 points. At this stage the Americans were the clear leader, with a margin of over a hundred contacts and the top overall multiplier number of 166, a slim advantage over their German neighbors.

Back at the end of the so-called entrance path to site 15W, where the apparent bicycle trail through the woods had been widened a bit for vehicles, the scene outside the competitor tent had settled into a routine. There wasn't much for the site managers, father and son Bennett, to do. In terms of visual or audible excitement, it was as quiet as any campsite out in the boondocks. The radio receiver remained on WATD-FM, where the announcer chattered continuously about the "World Radiosport Championship" now underway all around Boston. Most newspapers carried some form of feature on one or more of the visiting radiosporting personalities. Yet it would be hard to explain to a casual hiker or biker enjoying the state forest that there was a terrific battle underway, and that invisible and silent radio waves were crisscrossing the Atlantic not only into Europe but also into Africa—that these teams were making the best of those complex electromagnetic swords in a cosmic street fight of sorts.

As the referee, Wes Kosinski sat between Manfred on the left and Stefan on the right. No one had left his chair until Wes emerged from the tent for a cup of camp coffee and a nature break. "I think perhaps something for them to eat," Wes said in a combination statement and question. "Yes?"

This gave the Bennetts and me something to do, so we pulled together what appeared to be a tasty European-style lunch from the goodies Judy had brought from civilization: easy-to-nibble pieces of bread, Black Forest ham, several dark and heavy-duty-looking sausage sections, a smattering of vegetables, potato chips, and cookies. Wes filled a plate and took it back to the tent, and I carried the plates for the two competitors.

Manfred and Stefan were more than happy to get some food. I carefully handed them the paper plates and full plastic cups. There was little open space on the table. It was covered with gear and cables—as well as the finicky little scoreboard computer collector that had been so tricky and sensitive to get working. Amazingly, the SCC hadn't missed a beat ever since, faithfully updating the results back to headquarters and, therefore, to the world via the Internet. Each man somehow found a spot for the food and began to nibble at his lunch, all the while keeping the nearly twice-a-minute stream of contacts going. By this time they appeared to be in a nice operating rhythm, and I gave them a thumbs-up signal.

"Go ahead. Take a little listen," Wes told me as he somehow wedged his paper plate onto the table in front of his chair. He handed me his headset and immediately dug into his lunch. I stood just behind him and put on his headphones. Manfred was sitting to my left and was in my left ear, while Stefan to the right was the other half of the stereo audio. After the relaxed chitchatting in the bright sunlight by the Bennetts' campsite, and then the hurried food preparation for the team, I instantaneously entered a total change of emotional state, a new world. Manfred was operating on Morse code at a high rate of speed, about forty words per minute, while Stefan was on phone. Frankly, with both channels hitting me at once, it was hard to concentrate on either one. At Wes's listening position, he could switch entirely to either station with both ears, or he could separate them into a dual-stream cacophony.

Selecting to listen with both ears to Morse, I heard the sounds quickly transition into a high-speed coherent reality. When Manfred's station transmitted, the only thing I heard was a single tone of his code. But when he stood by to listen in the receive mode, an ever-changing olio, a hodgepodge of multiple tones competed in my ear. The first thing that hit me was how crystal clear the reception was. For people who are used to noisy urban settings where arcing electrical connections, motor vehicles, and faulty neon signs are the rule, receiving can be a constant battle to tease out signals into recognition—especially at marginal atmospheric conditions. Add in static crashes from a thunderstorm near or far, and . . . well, you get the picture.

Standing there in an already sweltering tent, out in that grouse-and-pheasant-hunting clearing, reachable only by bicycle or pack animal, or by our hotted-up Dodge Charger, that was as close to ham radio nirvana as it gets. When Manfred stood by, the radio world seemed to belong only to him. His frequency filters blocked out signals to the left and to the right of his setting. As I listened, it was like peering into the ionosphere in such a way that even if someone somewhere were using a single AAA-battery-powered teeny-weenie transmitter in Europe, they could be, and would be, heard with those earphones.

Usually the callers came in small waves: One operator might call with astonishing clarity, a Morse code signal all by itself—either weak or strong, it didn't matter since Manfred could hear it clearly (as could I) and go back to him right away. At times, two or even more would call in a little "jump ball" competition to see who could leap up with the strongest signal or otherwise stand out—say, with a different Morse speed to catch his attention. We all prefer certain frequency tones, consciously or subconsciously; at times, what appeared to me to be an obviously loud and clear signal was ignored by Manfred as he would go back to another caller, a signal that was tough for me to distinguish. Other times the opposite would occur, since I would hear people calling that he didn't answer. All in all, he was "swinging along right nicely" and racking 'em up at two per minute or more.

In my right ear, the voice operation by Stefan was, of course, different. For one thing, the receiver "width" was set much wider. Nearly all the intelligible sound we humans hear lies between 25 hertz, a low throb similar to a cat's purr, and 17,000 hertz. Thunder resonates around 40 hertz. The human voice ranges for the most part between 500 and 2,000 hertz, and a voice filter usually is around 2.5 kHz wide. While it makes for easy speech recognition, adjacent stations, or "aggressors" trying to winnow in closer and compete for a spot to operate, can be heard as well.

Stefan had a good frequency, even with his modest 100 watts and standard-issue antenna. It was impressive: Voices of the callers, either alone or in groups, popped up right out of the almost perfectly quiet background. The mandatory language of most contests, and of course the WRTC, was English. Yet the range of accents and dialects was both fascinating and challenging. Stefan spoke regular conversational English in a soft manner, unlike the more conventional Teutonic style of Manfred. His low-key enunciation was soft, with an added "sss" sound quite noticeable. However, on the radio he projected almost no sibilance, and even though you could detect a trace of German, his operating voice was clear as well as low in volume. That was a plus.

Some voice operators speak at a fever pitch. They yell, as if willing their voice to project acoustically all the way to the other station, rather than taking the atmospheric electromagnetic path. But Stefan wasn't that way. He kept up a reserved but highly effective rate. In fact, he seemed so intimately connected to his microphone, now on the swivel bar in front of his face, that it was difficult to hear him at all on the other side of the tent. It was clear that his style helped maintain less noise and preserve his energy for what was sure to be a long day and night. They had a full nineteen hours yet to go.

The operating style for both men was efficient, notable by a lack of wasted motion and energy. Their fingers flitted rapidly across the keyboards—almost too quick for the eye to follow. It brought to mind a magician's performance, in which "the hand is quicker than the eye." Only by listening to their respective audio channels—Manfred's distinct high-speed Morse code, and Stefan's quiet but crisp enunciation—could I follow and translate the minute keystrokes.

The computer has revolutionized radiosporting: Huge stacks of paper logbooks and other supporting records have been replaced. Now the digital wonder systematically records every single contact and knows whether a specific call sign is a new one or not. Individual keys are preprogrammed to send information that must be exchanged in every contact: call sign, brief signal report, location, and then a "thank you" at the end to confirm everything. In addition, the "I am seeking a contact" CQ calls are nearly always prerecorded.

---

As an example, a complete contest exchange, whether on voice or code, would go fast—something like this on voice. ("Phone" operators use phonetics, since a single letter would be difficult to distinguish.)

(transmit, prerecorded) CQ Contest Whiskey One Papa

(receive) Oscar Mike Four Foxtrot Zulu (A station from the Slovak Republic has answered.)

(transmit, live voice) Oscar Mike Four Foxtrot Zulu (you are) Five-Nine Eight

(The "5-9" indicates a strong signal, and the final "8" indicates zone number 8 of the world for the WRTC station.)

(receive) Thank You, Five-Nine Twenty-Eight

(This report indicates also the strong signal report, 5-9, and his zone number, 28, in Eastern Europe.)

(transmit, prerecorded) Thanks, Whiskey One Papa

That's it: a fast but complete contest exchange. The whole thing required about five seconds. Some contest exchanges could be longer, depending on the specific rules, with much more detailed information. But none would be much shorter than this.

After this brief exchange, one or more stations would call again—perhaps they just tuned across the frequency, or they had previously called but the operator at OM4FZ beat them out. If no one called, Stefan would push the "CQ" message key and repeat the process anew:

(transmit, prerecorded) CQ Contest Whiskey One Papa

Manfred Wolf
(DJ5MW)
operating
in daylight.
Photograph: Jim
George (N3BB).

Stefan von
Baltz (DL1IAO)
operating
in daylight.
Photograph: Jim
George (N3BB).

# Chapter 13

# AROUND THE CIRCUIT

One clear leader had emerged by the end of the fifth hour, at one o'clock in the afternoon. The much-talked-about American team at K1A had wrapped up almost 1,100 contacts, along with 166 different "multipliers," and the scoreboard now registered 693,000 points. The flawless technique of Dan Craig and Chris Hurlbut had been honed in many previous sessions operating together. They had a significant early lead over the next four teams, which were doing very well, grouped in a tight cluster between 562,000 and 578,000 points.

The early top-five challengers were English, Slovakian, German (Wolf and von Baltz), and a Ukrainian-Canadian duo. It was very early in the twenty-four-hour marathon, but these next four teams, on average, were already trailing the streaking Americans by a hundred contacts.

During these morning and midday hours, everyone maximized the two primary daylight frequencies. Because 28 MHz hadn't yet opened to any significant degree, these top teams had made a mere 3 percent of their contacts on that unpredictable band. Would the impish ionosphere of that frequency become a major factor or not? All the early leaders had focused on Morse code, as had been expected. Craig and Hurlbut had made only 11 percent of their contacts on voice, while the other teams ranged from a low of 4 percent (Slovakians) to nearly 18 percent (Wolf and von Baltz).

▷〜◊◊◊〜╟╟▫

The Brazilian-Bulgarian Friendship Team was tops among the sponsored teams, with an early lead after two hours. In third place after four hours, they had dropped to a (still-strong) seventh position at the five-hour mark. The Brazilian member, 27 years old, was an electrical engineer named Soni Leite (PY1NX). He had learned about Morse code and ham radio from his father, as had Atanas Koychev, also young at 40, who was his Bulgarian partner. Both men worked professionally as engineers and had extensive radiosporting experience in their respective countries. Leite already had one WRTC under his belt—from Brazil in 2006, when he was only 23!

The sponsorship of this team had a story behind it, as I suppose any commitment for $50,000 would. In previous WRTC events, donation levels for sponsored teams had ranged from $10,000 to $70,000. For the 2014 championship, the organizers had decided on four slots of $50,000 each. In May 2013, at the sprawling Dayton, Ohio, Hamvention (20,000–25,000 attendees, depending on who did the counting), the WRTC 2014 committee was closing in on netting a key sponsorship from a wealthy Brazilian industrialist named Atilano Oms (PY5EG, simply "Oms" to everyone), who was a very active amateur radio contester and DX (longer-distance contacts) enthusiast.

Oms had chaired the WRTC 2006 in southern Brazil, and he had been a significant supporter, as well as financial backer, of every recent WRTC competition. He had given an early verbal commitment of his support for the Boston-area event. Earlier in the evening, Oms had been socializing with "Krassy" Petkov, a longtime friend who had immigrated to the United States from Bulgaria in the 1990s. (Petkov was an excellent operator in his own right, and in fact he had qualified independently as a competitor representing the New England region in the WRTC 2014.) In the key meeting regarding financial support, Oms asked if Petkov could join them, and the committee agreed. At first the discussion didn't go well: The Brazilian economy had been slow to recover, and the currency exchange rate had deteriorated badly. The Brazilian apologized and said it would be very difficult to deliver on his promise of a team sponsorship. The committee expressed disappointment, and it was clear that Oms felt bad about the situation. Then he brightened, and as ever the deal maker, he turned to his friend and suggested that they cosponsor a team. It would be designated the "Brazilian-Bulgarian Friendship Team," with one operator from Brazil and one from Bulgaria. They would each cover half the cost. Petkov, who had founded a successful technology company in New England, said yes, and the committee readily (and with relief, I'm sure) accepted their commitment. Handshakes were made all around. Ultimately, Soni Leite from Brazil and Atanas Koychev from Bulgaria became teammates, although they had never operated together before and, in fact, had never even met in person.

⊳–ℓℓℓ–⊪

At the K1C site on the Fort Devens former military base, Ken Low and Scott Redd sat in the number-11 position after the initial period. Their plan was for Redd to "run" Morse code endlessly—just do that—and by 1 p.m. their nearly 1,000 contacts included only 10 percent on voice. "Keep the rate up" was their mantra. They kept their directional antenna on Europe 99 percent of the time, since a contact with a different continent was worth five points, considerably more than three points for most of the United States. In addition, they found that the American stations were easy to contact, even off the "back side" of the Eurocentric antenna. All of their equipment was working fine, and Low was very pleased with Redd's ability to sit there and "crank 'em out," as he described it, one after another on Morse code. The retired vice admiral was a driven man.

Another American team, Kevin Stockton and Steve London, had moved up to number 13 after being as low as 28. Stockton had qualified from the station that he and his father built on a mountain ridge in northwestern Arkansas. London had retired from his Colorado-based career in the Internet infrastructure business to a unique "round house" located in a spectacular setting in southwestern New Mexico. Both places were well known as amazing locations for competitive radiosporting, yet neither Arkansas nor New Mexico provided hands-on knowledge of New England radio signal conditions. So, it was an open question as to how well this southwestern team would do. Since Dan Craig and Chris Hurlbut, the "West Coasters" who were current leaders at K1A, seemed to have mastered the conditions in the Northeast, could Stockton and London do it as well? A subplot was developing: How would the "non–East Coast" American teams do in Boston?

Bouncing around, from ninth place all the way down to number 21 and then back up to 15, were Tom Georgens and Dave Mueller, located across the road from both the leading team at K1A and the fourth-place team at W1P (Wolf and von Baltz). Georgens and Mueller had done a bit more voice operation, at 20 percent, than the top groups.

The Russians (at N1F), who had won in Moscow in 2010, were struggling, although airplane noise was not a problem since they were hunkering down behind the quickly constructed plywood barriers. They kept on, but they had never liked the Mansfield Airport site. The team started at a decent number 14, and then moved up to seventh place after two hours of excellent runs on both main daylight bands. Then subpar third and fourth hours produced only 225 contacts between them, which was disappointing. They tried voice runs on 21 MHz, but the results just "didn't happen," and valuable time was lost. A fifth hour of 186 contacts helped, but some early damage had been done. They ended up after the first five hours in the number-23 position.

Sandy Räker and Irina Stieber kept up a solid performance on the grounds of the old state mental hospital. They improved steadily from a number-39 position after one hour, and by 1 p.m. had moved up to the twenty-fourth spot. All of their equipment continued to be problem free, and they were determined to do well to justify their highly competitive qualification placement in Region 2 of central Europe. Räker had made at least one good voice run so far on 21 MHz, but overall most of their contacts (82 percent) were on Morse, where Stieber was a world-class operator. Räker also was holding her own on code quite well.

The highly regarded Estonian team of Tonno Vahk and Toivo Hallikivi agreed with the Russians on one thing: They didn't like their operating site. Vahk felt they had called too many CQs without answers, and the constant streams of callers that were needed simply weren't materializing. He was also concerned about being so far inland—far away from the salt water that had a magical effect on signals (although several of his northern and inland neighbors were doing quite well indeed). Bottom line, the team started out at number 24, for them a prosaic position, and were number 35 after five hours. Their operating strategy was noticeably different, with a high number of voice contacts at 31 percent—much higher than most. On voice, the automated Skimmer Morse code detectors couldn't find them and splash their call sign and frequency on the Internet. Would that have impacted the number of times they were spotted?

The "Texan and Jersey Boy team," otherwise known as Team 59, George DeMontrond and John Crovelli, plugged away with their call sign, N1D. Their operating site at the old Fort Devens seemed to be working out, and they pursued their unique strategy with DeMontrond operating only with (loud) voice and Crovelli doing a great deal of "search and pounce" on Morse for both contacts and multipliers. In fact, after five hours they had one of the highest percentages of voice, and their 840 contacts included a strong multiplier total of 149. As a result, they had improved their standing by ten places, from an early number-49 spot to number 39, with 420,000 points.

The Youth Team was in there battling, from one of the state forest nooks, with all of their equipment operating just fine. Yet their relative inexperience showed, and they maintained a fairly consistent lower-quartile placement at number 47, with 801 contacts and 137 multipliers. They were, however, running ahead of the Japanese team, who were lagging with 716 contacts. While their SkookumLogger software continued to work well for them, their relative inexperience at this level of competition did not.

⊳⦿⦿⫴

How were the Lithuanians doing? Recall that they had lost their Lenovo laptop when a charging problem knocked it into "blue screen hell." The borrowed Dell had appeared to work normally, but then for some reason it faltered and stopped altogether once the contest started. Gedas Lucinskas, the team leader, had actually made a number of contacts using pen and paper, but he soon realized that this method was impossible. For one thing, it was simply too complex: He wouldn't be able to recall whom he had contacted previously, and who still was needed. For another, each team had to provide an electronic copy of every single contact to their referee within a few minutes of the end of the contest. So, even if somehow he were able to run up a big number of contacts on paper, there would have been no way to transcribe them into the required electronic format. It looked like he was toast!

One positive aspect of this, perhaps the only one, was that his teammate, Mindis Junka, was calm and running contacts smoothly from his station only one table away. Their antenna was working well. So, they still were in there with a chance, maybe—but if and only if Gedas could find a replacement computer, a system that would load the contest software program they were using *and* could be found in time to make a difference.

Their referee, John Laney, telephoned headquarters and requested help. "Will you authorize Gedas to replace his computer?" he asked.

"Of course" was the answer.

One of the site volunteers, new to the hobby and about to begin his freshman year at the University of Rhode Island, grabbed his cell phone and called a friend who was volunteering at another site. Good news: He had a laptop available. Bad news: It was at home. He rushed the PC to the Lithuanians' location (ironically, it was exactly the same model as the Lenovo T61 that had died in the hotel charging incident). Such teamwork and enthusiasm exemplified the wonderful volunteer spirit.

For more than two hours the Lithuanians had been dealing with a complete computer failure, and between 11:46 a.m. and 1 p.m. they were a one-radio WRTC team competing in a two-radio contest. They had fallen all the way to number 57 and then somehow had rallied back to 49 with only one radio.

Gedas must have been wondering if this would be the elusive third time, the "third time that never lies" of Lithuanian lore. His support guru loaded all the programs. They held their breath. The new system roared to life. It was as good as new.

He was back in action. Now they would see if they could fight back.

# Chapter 14

# SATURDAY AFTERNOON

The competition now settled into a Saturday afternoon routine. Each of the fifty-nine teams was on the air, and the early morning excitement had changed into an established pattern: Operate on the two main daylight frequency bands of 14 MHz and 21 MHz, but keep an ear out for the capricious 28-MHz band. It was the highest frequency of the five used in the championship, and if the ionospheric conditions hit a sweet spot, it could be the best and most productive of them all. In that event, every one of the teams surely would need to be right there on one of their radios. More likely, however, given the solar conditions, the band would open— if at all—in short little "burstlets" when the thin upper atmosphere of Earth, roughly 100 to 300 miles high, got just the right amount of solar energy. In that case, electrons would be knocked off the oxygen, hydrogen, and nitrogen molecules to float free and form a reflective shell. The best-organized and most-experienced teams would check every fifteen minutes—they had to be sure *not* to miss an opening.

The organizers had relatively little to do once the contest started. After they had confirmed that all teams were on the air, most of the risky start-up problems had passed. For a few glitches, such as the unusual computer failure at the Lithuanians' station, the site volunteers responded well. The organizers hoped the rest of the twenty-four hours would provide an even playing field, along with good radio signal conditions and no nearby storms. Along with the judges and several technical advisors, the organizers left their command center at the hotel and began to make the rounds. The weather was perfect, and every site received several visitors.

At the Germans' W1P site at the end of the long, stubby forest path, the Bennetts, Judy Attaya-Harris (a virtual knowledge-sponge regarding her new hobby), and I more or less took in the scene with small talk stacked on top of idle chatter. Hours slipped by. By Saturday afternoon, it was tempting to lighten the load, so to speak, of the food table, but we had to leave sufficient nutrients to keep Manfred and Stefan "hitting it hard," the little phrase I found myself repeating.

Occasionally I would lift the tent flap and "check on the boys" to see if anyone needed anything. The scene of silent tapping sounds (if they were on Morse) and calm, "soft talking" directly into the microphone (unlike some of the well-known screamers) concealed the mélange of signals from a disparate range of countries around the world. Michael Bennett's phone connected as our lifeline to the scoreboard. We were never far away from the updates.

Dave Sumner paid a second visit, making the rounds with two of the judging committee members. Later, Randy Thompson (K5ZD) , WRTC vice president for marketing and communications, also dropped by. Apparently the site's reputation as a food lover's paradise had spread. Thompson, still quite boyish in appearance with a mop of floppy hair, had lived in Massachusetts many years. A native Texan (he studied electrical engineering at the University of Texas), he had cut his radiosporting eyeteeth as a guest operator at the only contesting setup in Austin at the time. Thompson had lost neither his Texas accent nor his zest for life. He had competed in five (count 'em) previous WRTC competitions. He also admitted that between his day job in sales and marketing for a New England–based Internet software company, and a full year of nonstop planning and work for the WRTC itself, he hadn't kept his own station up to par. If all this weren't enough, he served as contest director for the highest-profile annual radiosporting event in the world, the CQ Worldwide DX contest. It's hard to fathom how he packed all this into one person. As if to underscore his frenetic nature, he was trying to make it out to half of the operating tents during the weekend.

While Thompson and I were chatting, another car slowly emerged from the forest, and another flamboyant radiosporting character got out. Ralph Bowen (N5RZ, nicknamed "the Gator" because he devoured the contacts in any radiosporting event) ambled over to say hello and check on the team. Gator, a lanky West Texan, along with his girlfriend—and both with accents to match Thompson's—had decided to "drop in" on New England and simply be a spectator. Our little Eden at the end of the world was becoming a popular place.

Insulated from the outside world, with the exception of the occasional roving radiosporting luminaries who peeked through the tent flap from time to time, Manfred and Stefan worked steadily. From a solid number-4 position after five hours, they had moved up to third place by 10 p.m. They had racked up over 1,900 contacts in ten hours, nearly 200 per hour, with a split of 80/20 on Morse and voice. The scoreboard indicated 1.73 million points. So far, the team had stayed on the two main daylight bands. Yet they had managed to get through to seventy-one stations on 28 MHz—apparently checking it often for openings—with a smattering of contacts every hour except the first one of the morning.

⊳─◦◦◦─╟⊩

Sites that might have contended for the "Most Visited" award, had it existed, included the German women and the Scott Redd team.

In addition, the sponsored duo of larger-than-life George DeMontrond and crusty John Crovelli were located at an easy-to-access spot that attracted six official reporters, including teams from broadcast and print media.

A reporter and photographer from *The Telegram*, based in Worcester, Massachusetts, had visited the site Friday, and they returned again when the contest was underway Saturday morning. DeMontrond was busy running a pile-up on phone (voice), so Crovelli took a break to field questions from the reporter.

"Why do you do this?" she asked him.

With his beard, wristband, and a hippie bandana, Crovelli provided snappy retorts in a distinctly New Jersey to-the-point fashion. He mentioned how their interest in radio had been a passion for both of them since they were teenagers. There was something magic about how a shortwave signal reflected at least once, often several times, off the ionosphere—it might have been coming from 10,000 miles away—and could sound fat and puffy at times, but gravelly and husky after it had been bandied about by strong magnetic fields in the polar regions.

Then he described the Houston-area automotive mega-dealer's commitment to the hobby—working it into a schedule crammed with business and civic responsibilities, as well as family time. DeMontrond had even spent time trying (unsuccessfully) to get the Olympic Games to Houston. With all his own radiosporting hours, Crovelli knew that the team would be serious and would prepare well, but since a top spot required both operators to be among the very best on Morse code, they would be a long shot.

"We probably have no chance to win," Crovelli said, giving up several minutes of lost contacts at his Morse station, "but we decided to work hard." Crovelli already had seen the power of Skimmer and knew that Morse callers far outnumbered voice due to the Internet spots. "Still, it's great fun."

One thing Crovelli wanted to mention in the interview was his and DeMontrond's appreciation of the site volunteers. Their designated site manager was an amateur radio operator from Massachusetts. He was joined by three Floridians and a Wyomingite—a group described by Crovelli as "team deluxe." He added, "They really took care of us." The volunteers provided stir-fried chicken Friday night, followed on Saturday by a sliced-steak dinner with potatoes and salad. Excellent sandwiches appeared at lunch. (Perhaps this site won the foodie award after all.)

Crovelli had operated for many years at a rented house on the Dutch island of Aruba in the Caribbean, and he had developed a taste for Polar Malta, a sugary nonalcoholic drink that originated in Venezuela as a by-product of the distillation process for beer. He had brought along plenty of the Malta. At any rate, the abundance of great food and continuous media attention on the grounds of Fort Devens made for a most interesting and calorie-rich experience. As for the contest, the well-cared-for team settled into the forty-ninth position after the first ten hours, with 1,601 contacts and 1.16 million points.

⊳⟋ℓℓℓ⎯╟╟

Perhaps the most national media attention was directed at Scott Redd. With his thirty-six years in the US Navy and his appointment as the first director of the US National Counterterrorism Center, that was understandable. His warm smile and neatly mustachioed countenance belied a steely commitment to excel. (His stated personal goal was to win the top prize of world champion as a single operator in each of the six annual major international radiosporting events.) The team's strategy was based on Redd focusing on Morse code runs—and that is what they did—so there would be no room for media breaks of any form during the contest itself. Even with their total concentration on the task at hand, they struggled: Their promising 11th position after five hours had dropped slowly to number 24 at the ten-hour point. Even so, their percentage of Morse code was over 90 percent, one of the highest.

▷〰〰▯▮

Sandy Räker and Irina Stieber settled into the task at hand. Their station was working so smoothly that mentor-advisor Tim Duffy left for a while to check on other sites. The team had lost a bit of ground, dropping to number 35 after ten hours. Their roughly 1,600 contacts yielded 1.32 million points, with a three-to-one ratio of Morse to voice. Nighttime conditions soon would approach, requiring even more focus on the code. How would they do?

▷〰〰▯▮

The early leaders, Craig and Hurlbut at K1A, had not only remained number 1, but had built a lead of more than 100,000 points with more than 2,000 contacts and 1.87 million points. The Americans were focused largely on Morse, with only 10 percent on voice. The Slovakians had eked out the second-place position, with a lead of only 30,000 points over Wolf and von Baltz; both these two teams had focused on Morse as well, but had been on voice a bit more than the leaders—making in the neighborhood of 20 percent of their total contacts up to that time with the microphone.

Tom Georgens and Dave Mueller had moved up solidly from number 15 to the 11th spot with around 1,800 contacts and 1.5 million points. Georgens intuitively felt that voice should be better—the exchange was so quick—and in fact the team had racked up nearly a third of their score on that mode. But even though they "pushed voice," the magic conditions for a low-power voice run just weren't there. The contacts simply didn't come fast enough. Besides, the Morse code presence of every team was uploaded instantly to the entire world, thanks to the automated Morse-detection Skimmer software. In this contest, Morse ruled.

⊳‿ℓℓℓ‿╫╟

One of the biggest upward movements in the afternoon period on Saturday was accomplished by the American team of Kevin Stockton and Steve London. At the five-hour point they were positioned at number 13. It had been a difficult start for them—they had been twenty-eighth after two hours—but they hit their stride and moved all the way up to sixth place after the tenth hour, with nearly 1.6 million points. They were making loads of contacts and had serious momentum, yet their "multipliers" were a bit lower than the top teams'. If they could close that gap, they had a chance—a real chance.

⊳‿ℓℓℓ‿╫╟

As dusk approached to mark the end of Saturday's full-daylight period, the Russian gold medalists and their Estonian runners-up were less than pleased. The Russians had slipped a bit farther, from 23 to 26, even with their laser-like focus on Morse code (86 percent). They had done well on multipliers, however, with a total equal to, or exceeding, half the teams in the top five. But they were running behind on contacts, with 1,700—nearly 20 percent lower than the leaders.

The Estonians were struggling. They had made 1,565 contacts in total, and in the past five hours only one operator was able to achieve one hundred-contact hour. In other words, for these ten "operator hours," the team had cracked the key hundred-per-hour rate at one of the two stations only for a single hour. Tonno Vahk and Toivo Hallikivi were demonstrably world-class operators. Yet, with only one hour of a hundred contacts or more, their performance paled compared to the Americans at the K1A station, who knocked off *four* hundred-plus hours in the same time, and to the German men, with *six*.

Vahk sensed something was wrong, and more and more he felt that the slight hillock of ten feet in the direction of Europe was hurting them. But was it? While Hallikivi had operated mostly on Morse, Vahk had spent an unusually large portion of time on voice, without the tailwind of those Skimmer spots. Even though Vahk had begun to focus more on code as well, the team's overall voice contacts were still greater than 27 percent after ten hours. The team dropped even farther off the pace to number 42, down from 35.

⊳⟍ℓℓℓ⟋╫�╫

On the other hand, Filippo Vairo and Paul Whitman, the Youth Team, were going full-bore now with 1,621 contacts. This exceeded the much more experienced Estonians, and the young men had pulled up from number 47 all the way to number 32.

The Japanese team of Don Kondou and Hajime Hazuki remained far behind the leaders at number 56, even with one of the very lowest percentages of contacts on voice, at 9 percent. Perhaps the fast contest English was difficult for them. The two men had traveled a great distance and had brought fourteen significant pieces of luggage (with an overweight charge of $700) all the way from central Japan to Boston. The array of equipment had barely fit into their driver's car, even with cramming. They were up against elite competition, but they never stopped trying their best.

⊳⟍ℓℓℓ⟋╫�╫

The sun was moving over the treetops to the western horizon now. Conditions soon would change dramatically, moving from daylight to darkness, and this would require new strategies and possibly different skill sets. As the dusk deepened, one team in particular felt encouraged. The Lithuanians had been able to get both operators going full speed at 1:10 p.m., after more than five hours of chaos. Now, with the replacement laptop, both Lucinskas and Jukna were going strong. Finally, everything was working the way it should. After falling all the way to number 57 after two hours, they somehow had scratched their way back, with only one radio, to number 50 by the fifth hour. With both stations running, they stormed all the way up to twenty-first position after ten hours, with 1,820 contacts and 1.45 million points. It was an impressive feat.

Walt Marshall (W7SE) preparing supper at the DeMontrond–Crovelli site. Photograph: George Wagner (K5KG).

# Chapter 15

# DUSK

Radio waves are electromagnetic bits of energy. Depending on the frequency of their sine-wave-like motion as they travel in dynamic form at the speed of light, they exhibit different properties. That is why we receive commercial FM radio signals only as far as what is called "line of sight," or directly from the transmitting tower to our car or home. FM signals don't, as a general rule, fly up and bounce back to Earth from some reflective surface. AM radio waves, on the other hand, are much slower than FM in terms of how fast their little sine waves jiggle. As an example, commercial AM radio is transmitted at frequencies of around a million cycles per second (1 MHz, in technical terms), while FM radio stations use hundred-million cycles per second, or 100 MHz—a hundred times faster. These two frequencies don't act the same in the ionosphere.

AM signals follow Earth's curvature with what are known as "ground waves" during the day, because the part of the signal that flies up into the ionosphere is literally gobbled up, is completely absorbed. This upper-atmospheric layer of hungry, super-absorptive molecules is called the "D Layer," and it exists only during daylight. At night, without the solar energy boost, it dissipates almost completely as daylight transitions to dusk and then to full darkness. Freed from its "D Layer chains," the AM signal radiates upward and bounces back to Earth off a higher band of energy formed by other types of radiation. Coupled with the curvature of the planet and the ionospheric energy bands in the atmosphere, the signals return to Earth hundreds or even thousands of miles away. This explains why powerful AM radio stations in New York or Atlanta are loud and clear in Kentucky at night, and why in the 1960s kids living in a vast twenty-state area got their rock 'n' roll kicks listening to WLS in Chicago.

Amateur radio "bands" exhibit quite different propagation characteristics, depending on their speed of oscillation, or frequency. Some are "nighttime" bands. During the day these signals radiate upward and reflect back to Earth within hundreds of miles, but at night they extend out thousands—sometimes many thousands—of miles by reflecting off a higher "mirror."

Other bands (and I am oversimplifying here) are "daylight" bands and reflect—like good little soldiers during the day—off higher-altitude free electrons. This results in distances that range from many hundreds to tens of thousands of miles. At night, however, they lose their ionospheric mirror and don't return to Earth at all. Depending on the amount of sunlight over the signal path, these pathways can open a bit, or they can be completely "dead," with no signals. Of course, things aren't always so simple. The twilight periods of dawn and dusk combine both day and night characteristics and sometimes produce extremely interesting conditions that feature the best of both simultaneously.

Signals on the daylight bands usually are clear, since atmospheric noise is low. An exception occurs when the radio waves must travel over the highly magnetic polar region. Yet generally it's easy to hear signals during the day. On the other hand, nighttime bands can be more susceptible to lower-atmospheric phenomena, including lightning-induced static and other disturbances. In terms of the WRTC teams, the overall skill sets required for day and night operating were similar, but they differed in important details. Overnight conditions could produce significant changes in the team standings. A loose analogy might be made with race car drivers who alternate between dry-track and wet-track conditions, between rain and clear weather.

The WRTC teams were spread out in an arc west of Boston, from the New Hampshire border almost to the border with Rhode Island, then east nearly to the land bridges to Cape Cod. Although the official sunset time Saturday was around 8:22 p.m., dusk in the summer began well before that time and extended well into the evenings. To the east, in the direction toward Europe or Africa, it was already dark. Ionospheric conditions were changing dynamically.

Twilight lingered at WRTC site 15W. Visitors had come and gone. Now, with darkness inching through the forest tree by tree, calls from birds and squirrels transitioned to sounds of owls and hawks looking for their prey. The radio conditions at the campsite changed to track the ionosphere and its different layers of absorptive molecules and free-floating electrons.

By sundown, Manfred Wolf and Stefan von Baltz had completed two of their very best hours—a total of 518 contacts altogether between 6 and 8 p.m. During that time, 14 MHz, a key daylight band, was still fueled with sufficient solar energy. The signal path between New England and Europe (with very long-lasting daylight that far north) remained fully "open." The primary nighttime band, 7 MHz, now was getting the needed reduction in the sun's rays to find its sweet spot. For a two-hour period, both of these bands were in extremely good shape. The two operators were on a roll.

Both Manfred and Stefan were extremely busy, but from time to time they managed to pick at the food I had brought into the tent. Wes signaled for me to take his place for a few minutes while he ducked out to get some fresh air and eat a bite. Listening to Manfred's radio on the 7-MHz band was an audible form of peering into a brilliantly clear ocean of beautiful Morse code. Senses of colors and sounds began to mix. I could hear signals from other stations, yet the filters in the receiver were cleaving a wondrous, narrow opening. Callers from Europe, as well as from the eastern and midwestern United States, appeared with astonishing clarity. Every signal was just a little different: Some were faster, others slower, some with higher timbre, others with lower. In my mind, a symphony of colors appeared in addition to the sounds: Higher frequencies had a cooler blue image, while the stations with a basso sound triggered orange or even red. The hues flashed and danced to the beat of the Morse code. This was radiosporting at its most exhilarating, and I couldn't help but want to sit in that chair and "run" those Europeans with the magical band conditions.

*Chapter 16*

# OVERNIGHT

~~~

The leading team, Craig and Hurlbut at K1A, or "Team One-A" as some called them, continued to pull away. They were firmly in the top spot and appeared to be growing stronger every hour. They now had more than 4,200 contacts. Overnight they had increased their margin from 110,000 to 530,000 points. The primary nighttime band was 7 MHz, and they didn't miss a beat—notching 985 contacts by 4 a.m. The leaders were doing very well at nearly everything, including the critical multipliers. They had 400 "mults," a significant advantage over Wolf and von Baltz, who were in second place with 386.

The Americans utilized a simple and effective technique to add multipliers: If someone in a new place called either of the two operators, they would complete the contact on that frequency, and then ask that operator to move to another band, where they had not yet gotten that specific country. One of the teammates would leave his "run frequency" and contact the cooperative operator at the "move-to frequency." After each additional contact, the men at K1A might request that the operator at the other station move *again*, where he was needed as well. Occasionally, when conditions were right, astute radiosport operators could end up with three (rarely, and deliciously, even four!) new multipliers from a single station by using this procedure, called "moving the multiplier." Craig and Hurlbut actually had several of these four-band moves. It was like hitting a home run—a four-bagger. They covered all the bases.

Most of the time, their original frequency remained open. But not always. They risked leaving a nice, comfortable, established calling frequency to move to another (even for a few seconds), and someone else could always come in and poach—grab possession of the original spot. Usually, almost always, they could reestablish themselves somewhere on a new little piece of the original band. All in all, they were "masters of the mults."

The Slovakian team operated at a spot officially designated as 14A, Freetown Breakneck Hill. The place was very near the border with Rhode Island, to the southwest of Boston, and their location seemed to be favorable. They had been doing well the entire contest: in third place after five hours, second after ten hours, and fourth after fifteen hours. Now, just before first light in the east, they were back in second place at 4 a.m., the twenty-hour point. With only four hours to go, the top five teams were separating themselves from the others.

In the battle for the remainder of the top echelon, the Slovakians led by a nose (only 20,000 points) over an American-Canadian team of Fred Kleber and John Sluymer. Kleber, a wireless industry consultant, had lived in the US Virgin Islands for the past five years, and had qualified from the Caribbean regional zone. Sluymer, a senior project manager in broadcast infrastructure, had built an amateur radio superstation in Ontario and was well known in North American radiosporting. He had been a gold medalist in Brazil in 2006, so clearly he was an elite operator as well as a topflight engineer. They had moved up in the ranking from the number-12 position at five hours, to fourth at ten hours, and had now locked in at a solid third-place position overnight.

The final top two teams, in a tight grouping at fourth and fifth, were also separated by only a small margin. The hard-charging American team of Kevin Stockton and Steve London, and Manfred Wolf and Stefan von Baltz, were only 20,000 points apart, at 5.37 million to 5.35 million. The Germans had maintained a strong position nearly throughout the competition. Stockton and London, however, were moving up sharply, having jumped from as low as twenty-eighth to thirteenth, then to sixth, and on to fifth place five hours later. Now, at the twenty-hour point in the competition, they had moved into the fourth position. One key reason for their amazing headway was their success on 7 MHz: They had made more contacts on that band than any other team. London usually operated by himself from his home at a spectacular location. He said, "Forty meters [7 MHz] on Saturday night to Europe was better than I've ever heard from New Mexico, even with a [huge] four-element antenna. It was fantastic!" He and Stockton were experiencing all this with a simple wire dipole and a modest 100 watts. Viva New England!

It was well known in radiosporting circles that London was a driven man. He hadn't left his chair at all up to this point, and he never would until the contest ended. He commented later that the temperature was hot during the day on Saturday: "There was no Mylar reflective cover on our [tent] top. We sweated, and I never had to pee as a result. We never had any idea how we were doing." They just kept at it, like all the other teams.

Stockton and London switched operating positions between 7 MHz, the nighttime band that had turned in—if the ionosphere can be described that way—a once-in-a-lifetime performance, and 14 MHz, usually a workhorse daytime band. Stockton had sat down to do the final hours on 7 MHz, and London took on the task of milking the last spurt of activity that had emerged as Europeans awoke and many radio amateurs turned on their radios.

Not every radio amateur is a dedicated radiosport enthusiast, by any means. In fact, many operators had listened early Sunday morning and had no idea in advance that there was a contest going on. They certainly would have heard the rat-a-tat-tat of fast Morse code and the frantic calling on voice, depending on their preference. Most would have recognized that a contest of some sort was underway. They could exchange reports quickly with a few stations who appeared to be serious, and whose signal was easy to copy.

At thirty minutes past midnight, Stockton was operating on 7 MHz. Its nighttime (dark) path now extended across the Atlantic only into the western portions of Europe. Many operators in North America and South America were asleep. As a result, the contact rate was lower, but still active.

Steve London, meanwhile, was milking the last vestige of 14 MHz to Europe, and the "breakfast crowd" provided a fresh supply of contacts. He was pleasantly surprised by the bump in activity, since a continuous stream of stations called. At 12:31 a.m., he sorted through an especially unruly gaggle of competing signals. At first, the Morse characters for $S_5$ stood out, but then the signal stopped abruptly. London then could hear part of another call and sent a question mark. Immediately, UR5E, a Ukrainian station, popped to the top of the callers. Both London and the European operator were sending at a rapid speed, in the area of forty words per minute; the exchange was quick. London stood by once again.

This time the previous caller came through clearly from the start, sending at a moderate pace. Apparently, the operator previously had sent $S_5$ instead of $SV_5$. The $SV_5$ was followed by a "portable" indication, and then British call letters. This sequence indicated that the operator was licensed in the United Kingdom, but now was located in a portion of Greece. That was not unusual for someone on holiday, or spending time in a rental unit.

The prefix *SV* is the usual designation for Greece, but several of the Greek outer islands are located sufficiently far from the mainland that they qualify as separate "countries" for the purpose of different "radio lands," in much the same way as Hawaii and Alaska count as separate "countries" from the lower forty-eight American states.

The SV5 prefix signified the Dodecanese, an island group in the eastern portion of the Aegean Sea. These islands include Rhodes as well as multiple small tracts of land near the Turkish mainland.

The time was 12:32 a.m. in the WRTC tent, which corresponded to 7:32 a.m. on Lipsi, an islet of seven square miles, with a population of 790 people (2011 census). Perhaps the operator on the other end of the contact had just now switched on his radio, and was having his first cup of coffee. Neither Stockton nor London had heard a station from the Dodecanese yet. In any event, the caller was a welcome addition to their logbook. The exchange went smoothly, and a new multiplier went into the books. That certainly spiced up the W1Z team's spirit as the contest moved into the dreaded very late overnight hours.

⊳–꜀ꞁꞁꞁ–ⲓⲓⲓ

Wolf and von Baltz were in the number-5 position now, trailing Stockton and London. The Germans definitely were still in the mix for a top position. They had been near the top with rankings of fourth, third, second, and now—by a whisker—number 5 with only a few hours remaining. If they had been able to see the scoreboard results—and of course they couldn't—they would have known that Stockton and London had beaten them by over 150 contacts on the nighttime bands. They did manage, however, to maintain a slight lead in the multiplier totals over the Americans. All in all, the four teams clustered between the second and fifth places were so close that it would be a sprint to the finish from 4 a.m. until the end of the WRTC at 8 a.m., only four hours later.

Another team making a sharp run-up in the standings included the Lithuanians at W1A. Gedas Lucinskas had overcome his dead computer at 1:10 p.m. Saturday, when they finally got back on the air with both stations. The team had dropped all the way back to the number-50 spot shortly before that but then started coming back. By 6 p.m. they had moved up to number 21. They cracked the top ten after fifteen hours, at 11 p.m. Overnight, using both nighttime bands, they made 1,317 contacts, a strong showing. By four in the morning, their score was nearly 5 million points, and they held the ninth-place spot.

When the computer failed, their referee, John Laney, had called in all the help he could. Once the replacement Lenovo laptop arrived and was online, he had little to do but listen to both operators on his stereo headphones and offer quiet support. Laney, the federal judge who specialized in bankruptcy hearings, had recently accepted another ten-year appointment to the bench, an extension that would take him beyond 80 years of age. He was renowned for his photographic memory, and from time to time he amazed people by recalling a radiosporting event from twenty years ago, perhaps on a Caribbean island. He would be able to remember the time and place, and even whether a particular call sign was one of 3,000 or 4,000 he notched in that event! No one I knew had that sort of ability to recall radio call-sign minutiae. The man was a savant.

On the other hand—and maybe this was one aspect of his mutant-like memory ability for specific things—Laney had struggled for many years with a mental block concerning the basic technical theory and electronic requirements that were required to pass the higher levels of amateur radio licenses. For over five decades he had held a basic General Class amateur radio license, K4BAI, based on a test he took (and narrowly passed) as a teenager in Georgia. He had tried again and again over the years to achieve the 74 percent score needed to pass the more extensive Extra Class license that enabled the holder to operate on additional, more desirable, frequencies.

Finally, a few years ago, a friend suggested new software, an online program "guaranteed" to get Laney over the hump. In his words, "They said that if I could answer their software questions, I would be able to pass my Extra Class test." He studied "like a driven man" during breaks in his caseload and in other free moments. Then, after well over fifty years as an active ham enthusiast, and with dedicated cramming, one day the online program flashed a message: "You are ready to schedule your license test." He was truly exhilarated when he squeezed by and passed the actual government test. He had made it! Now Laney was at the WRTC watching "his team" battle back.

▷ ℓℓℓ ⊩

In identical tents that were now cooling off after a broiling Massachusetts summer day, every team had its own story. Tom Georgens's sponsored team was number 16 at 11 p.m. and remained in that same rank after twenty hours, with nearly 4.8 million points. Georgens and Mueller had over 3,700 contacts, far below the 4,200-plus of the leaders, but they had racked up 371 multipliers. They were one spot ahead of the defending champions, who were 30,000 points behind.

The Russians were nearly 300 contacts behind Georgens and Mueller. Yet they had a few more multipliers, as well as more contacts with Europe, which counted as five points per contact compared with (mostly) three points for US stations. Grinding away at the Mansfield Airport, they also had been moving up through the ranks—from the mid-20s in the afternoon, then to number 18— and now were seventeenth place in the early morning prior to first light. According to Denis Pochuev, their Russian-speaking referee, the team had settled down after a rough start. They didn't show any obvious frustration and were demonstrating "beautiful teamwork" running Morse.

At the NIA station, Sandy Räker and Irina Stieber had settled in nicely. All their equipment had continued to work perfectly. In Sandy's words, "We didn't have any problems. No line noise. No splattering. Nothing." They felt especially comfortable with Rusty Epps, their referee. The women focused totally. Neither one ever left her chair—no bathroom trips outside to the portable toilet, no stretching, no breaks at all. They were in the twenty-third position at 4 a.m., with 3,476 contacts and 362 multipliers. Overnight they had made 1,036 contacts on the two nighttime bands. With only four hours to go, they had earned 4.61 million points. A position in the top half in their inaugural competition against the world's best would be a very solid finish.

Ken Low and Scott Redd were positioned at number 19. Their strategy of having Redd focus entirely on "running" stations was working, and they racked up 1,269 nighttime band contacts out of their total of nearly 3,800. They had staked out two terrific frequencies—positioning themselves in prime territory—precisely at the very lowest portions of the 7-MHz and 14-MHz bands. These were locations where their 100-watt signals were relatively free of intruding stations, and they could hear callers easily. Unfortunately, a subpar multiplier total of 350 limited their score a bit, at 4.7 million, compared with what it could have been.

Those two great "run frequencies" might have been a boon and a curse at the same time: They were so valuable and attractive that the men hesitated to leave them. As a result, they contacted only the stations that called them at these two spots. They didn't go looking for unusual stations in new countries. Most of their callers were from the usual populous countries in Europe and North America. Few, if any, new places called in; there was a smaller number of new multipliers as a result. They also missed a key tool used by the leaders at K1A: If a station called from a location (a multiplier) they needed on a different band, Redd and Low didn't ask that operator to move, since they were concerned that their valuable real estate (in terms of the radio spectrum) would be lost in a second to other operators, who would swoop in and take it over. While they were making new contacts at a very solid rate, they weren't leveraging these numbers with the additional "mults." Looking back, Ken Low acknowledged that this was a flaw in their tactics.

On down in the pack, but making progress, was the Estonian team at N1T. They had improved their position a bit, up to thirty-first place in the overnight hours, buoyed with a robust multiplier total of 373. But their contacts were well off the pace, at around 3,400. Their results in the daylight on Saturday on both the 14-MHz and 21-MHz bands definitely appeared subpar for a team with their past track record. In fact, their nighttime band numbers also trailed others. Tonno Vahk was definitely unhappy and believed that their location simply "didn't get out to Europe" as well as other sites. He was still of the opinion that the slope of the ground, very slightly upward in the direction of Europe, was a limiter for them.

Three other teams were struggling as well. John Barcroft and David Hodge at K1W were located to the north, near New Hampshire. Their 14 MHz contacts totaled over 1,600—a full 200 higher than the Estonians—however, they had a fairly anemic multiplier number of only 341, which was well off the Estonian team's number. Overall, Barcroft and Hodge were holding the number-36 position with a score of 4.28 million.

The Texas–New Jersey–sponsored team of DeMontrond and Crovelli had continued in the mid-40s but then dropped a tad to number 47 at 11 p.m. Although it certainly wasn't a prime reason for their ranking, they were bothered overnight by a nearby team's "key clicks," a form of interference that made listening difficult. Like others, they discovered that 14 MHz was productive. It was "open all night over the North Pole," in their words. As a result, DeMontrond piled up prodigious numbers of voice contacts, even though the signals were garbled and hard to understand with the polar flutter.

By 4 a.m., they had made 1,400 voice contacts out of 3,312, for an unusually high 44 percent of their total. They felt that their overnight performance had been strong and asked their referee how they were doing.

"Not in the top half" was his response.

But they "never quit and stayed in the chair," according to Crovelli, forestalling even a single potty break while the contest was underway.

The Japanese team of Kondou and Hazuki ended the overnight ten-hour stretch with a bit over 3,000 contacts and 3.13 million points. They stayed on Morse for the most part and had one of the lowest voice contact percentages at 11 percent. But they simply didn't make as many contacts as the other teams, and they were especially down on multipliers, with only 309. On the other hand, they had one of the highest numbers of contacts on the 3.5–3.8-MHz band, with nearly 500. Most of these, however, were with North American operators and counted only three points per contact.

⊳〜ℓℓℓ〜╫

My day ended at 11 p.m. when I said good night to Michael and Derek Bennett and waved a quick bye to the three men inside the tent. Inwardly, as I approached the car in the darkness, little pangs of "How do I get back to the hotel?" bounced in my head, but fortunately the trip was now routine. Through the black forest at five miles per hour, left on Cuttersfield Road for a mile. Fortunately, all the night moves were the right moves this time.

The very welcome visage of the hotel appeared at 12:30 a.m. I pulled into a not-very-close parking spot, simply exhausted, and glad to be back with nary a wrong turn. The lobby bar crowd was plentiful and still going strong. But I bypassed all that and felt half-way decent when the alarm went off at 4:30 Sunday morning.

L to R: Jozef Lang (OM3GI), referee Charles Wooten (NF4A), and Rastislav Hrnko (OM3BH). Photograph: Bob Wilson (N6TV).

Manfred Wolf (DJ5MW) operating at night. Photograph: Wieslaw (Wes) Kosinski (SP4Z).

Stefan von Baltz (DL1IAO) operating at night. Photograph: Wieslaw (Wes) Kosinski (SP4Z).

Referee John Laney with the Lithuanian team at W1A during WRTC operation. Photograph: Bob Wilson (N6TV).

The German women operating at night. Photograph: Steve Moynihan (W3SM).

*Chapter 17*

# MORSE MAGIC

Speaking in the human voice seems natural as a way to communicate. Of course other methods, including Morse code and radio teletype, exist as well. In terms of efficiency, these techniques, as well as more current digital protocols, far surpass voice in their ability to transfer information under difficult circumstances. These modern digital wonders use statistical probability methods to succeed when the human ear hears absolutely nothing at all. For many radio operators, however, Morse code represents more than simply one additional way to transmit and receive information. For them, Morse seems to bridge an emotional connection of some sort. The sounds of the characters invoke a deep-seated resonance, an auditory-sensory connection to the human spirit.

In World War II, a BBC news announcer, 36-year-old Douglas Ritchie, assumed the on-air pseudonym of "Colonel Britton." He noticed that three staccato notes, followed by a big, long, booming tone at the start of Beethoven's Fifth Symphony, echoed the Morse code for the letter *V*.

Imagine the four sounds now:

DUNN DUNN DUNN DUUUUNNNNNN

Now a bit lower in frequency:

**DUNN DUNN DUNN DUUUUNNNNNN**

The British government used this dramatic opening for the BBC news beginning in July 1941, and continued for the rest of the war. The letter *V* also stood for victory (in English), and the connection was expanded to an extensive propaganda campaign that encouraged listeners all over the United Kingdom, as well as continental Europe—much of which was occupied by the Nazi military—to write or scratch the letter whenever possible to indicate resistance and hope, when hope sometimes seemed to be a long shot.

As a modern extension to this story, the great "arena band" Rush transformed the Canadian airport code for the Toronto International Airport into a hit song. Someone knew Morse code, because the airport designator code became a dramatic opening for the band's hit "YYZ."

The Morse characters for these letters

— • — —     — • — —     — — • •

became a virtuoso drum performance when the sounds were combined (without spacing) multiple times as the lead-in to the song. Several YouTube performances exist, including a live show in front of 70,000 people in Brazil.

Fascinating linkages of music and speech to human emotion exist. Several of these clearly are based on Morse code.

Morse code is different from the voice mode, or phone operation, in amateur radio. The mental decoding process needed for Morse is complex: For some, it presents no problem, while for others, code skill seems impossible to learn. Few people reach elite levels of capability, although with practice, most can be proficient at a basic level. Low correlations have repeatedly been found between Morse code achievement and intelligence, educational level, mechanical ability, and knowledge of subject matter.

The code can have its own dialects or accents, similar to variations of speech. In the early days of radio, operators sent by hand in one way or another—using pure hand keys—with only one spring and a little finger pad that he (it was nearly always a man back then) pressed down to make the connection. A short dit was just a fraction of a second, and a long dah was at least four or five times that. Personalities soon developed, with accentuated dits (even shorter) and long dahs (even longer). This style was known as a "swing," and the epicenter of this funky Morse personality was the American Great Lakes region. To have a "Lake Erie Swing" was either a badge of honor or a snarky criticism by Morse purists, depending on your artistic point of view.

Morse keys were developed with elaborate mechanical springs and contraptions to make the dits automatically. These ranged from plain Jane, quotidian, stripped-down models to finely tuned mechanisms with high-quality coatings and world-class metal work. Several commercial companies emerged, and one, the Vibroplex Company, is now over a hundred years old (with a new, youthful owner). Thousands of the company's faithful (may I say obsessed?) customer-adherents remain committed to the art of preserving Morse "the way it should be." Each operates with his or her unique "fist," the word that describes a person's code personality.

Yet things change, as they always do. And just as the mechanical age transitioned to the electromechanical, the era of "bugs," as the Vibroplex and its cousins were called, reshaped itself as well. New electronic technology generated dits and dahs automatically, and the operator no longer had to speed up or slow down based on how fast his (or her) wrist and fingers moved. Now they could simply turn a single knob to adjust the speed from a few words per minute to an ear-and-brain-numbing blur. In addition, the electronic circuit allowed the dits and dahs to vary (ditzy dahs . . . I just couldn't resist) in relation to each other—it's called weighting—so there *was* some personality possible, even with this newfangled mechanism. While the Lake Erie Swingers and other more hidebound Morse enthusiasts stayed with their bugs—and some do still to this day—most radio operators moved to electronic keyers, as they became known. You operate this thing by touching your finger and thumb to a mechanism to make the dits and dahs. It's known as a "paddle" and, as could be expected, paddles have evolved from basic electrical switches to elegant bejeweled swivels with magnificent machining and plated metals, topped with gorgeous high-quality plastics. The contact precision can be adjusted from whisper-light to clicky-clunky. Some modern paddles are electromechanical objets d'art. Morse code has not only survived but has also grown, helped along by the sophistication and variety of the gadgets by which it can be sent.

As a hyper-fueled marketeer might say, "But wait, there's more!" The computer itself can be programmed to send code. Software programs soon appeared to translate a keystroke into a perfectly formed Morse character. Modern technology never sleeps. In a world-class event like the WRTC, every team used computer-generated code, either from the keyboard or from previously stored messages. In addition, nearly every operator had a paddle to send manually through an electronic circuit if required. Some competitors actually preferred to send by hand rather than from the keyboard. Usually these were old-school, two-finger hunt-and-peckers, although in John Crovelli's case he was a one-finger ultra-traditionalist.

As I sit here attempting to create some type of image to translate the magic of Morse code, three musical performances come to mind—sounds that absolutely floored me when I first heard them. They are the Beatles's "I Want to Hold Your Hand" (which I first heard on the AM car radio, driving back to college one night); Wagner's "Ride of the Valkyries" (which shook me to my core early one Sunday morning on nonstereo FM back in the day); and the incredible, dramatic percussion introduction to "Money for Nothing" (a.k.a. "I want my MTV") by Dire Straits. If it makes any sense at all, good Morse code gives me the same mind-tingling thrill.

# Chapter 18

# TREASURE HUNT

Imagine you're a video gamer. You're very good at it. You made it through a tough preliminary qualification, and now you're competing in the final round of a major international tournament. Your opponents are scattered all over the world, yet the apparent reality is that all of you are together on one battlefield. The competition—to be crowned the best gamer in the world—lasts for a grueling twenty-four hours.

It's 4 a.m. You've focused on sights and sounds for twenty hours straight. Sunrise is still an hour away. Your joints ache, but you don't dare get up. Even for a minute. One "kill" might mean your toughest challenger pulling ahead of you.

It's been a long journey. The games sometimes end with a surprise—a hidden treasure exposed. Can you find it? Will you be "the one"? You can taste victory through the fog of fatigue.

What took place at the WRTC 2014 really happened, but it could have been scripted by a video game developer. A hidden treasure *was* exposed. Some teams found it; others did not.

Together, my alarm and the hotel telephone created a cacophonous jangle. Going back to sleep wasn't an option. This morning, however, I didn't have to deal with the hangover of the Wrong Way Curse, wonder if Stefan von Baltz and Manfred Wolf would get any sleep, or worry over Stefan's audio-isolation idea working. After a quick shower, I padded silently through the hall, walked down the stairs into the lobby—a much more serene place early Sunday morning—and then out to the car to head back south.

The red Charger cruised along now in the pre-dawn calm. First light had preceded the 5:20 a.m. sunrise by an hour. The trip was beginning to look familiar, and the weathered sign and arrow pointing to Camp Squanto confirmed the turn onto Cuttersfield Road.

It was 6:30 when I parked the car. The Bennetts (did they ever sleep?) sat outside their tent. Coffee was at the ready. Inside the operating tent it was quiet. Only a hint of mechanical clicking provided any sound: Both Manfred and Stefan were on Morse code. I brought Stefan and Wes some coffee—Manfred was a cola drinker, but waved that option off. With less than ninety minutes to go, clear signs of exhaustion were checked by a mixture of adrenaline and determination. These two men had been *so close* in the WRTC 2000 in Slovenia, where they failed to medal by a miniscule 1,050 points—roughly one-tenth of one percent. Based on their expressions, it appeared they *really* didn't want to miss the top three once again.

There it is—the treasure. Your screen explodes with hideous ogres—massive misfits with horns, beaks, and thunderous screams. You, the brave gamer, remain strong. You dodge the slicing swords and lacerating laser rays. A spray of automatic weapon fire somehow misses you. Then you attack. Pixels scatter in sharpest high-definition video. Your headphones pulse with eardrum-shattering audio. You move forward. Toward the prize.

▷ ℓℓℓ ▮▮

In the reality of radiosporting, there are times when surreality exists: to experience the bands explode with a great "run" in a contest; to stumble on a breakthrough multiplier. In this case, the hidden treasure was more subtle. Yet, at the end of a long challenge, in its own way it would be every bit as exciting as the *Sturm und Drang* of video game battles.

▷ ℓℓℓ ▮▮

It was barely detectable now. The signal came from the right direction—from Europe. The voice now peaked up just a bit out of the noise. The call sign indicated a headquarters station. The installation probably was well equipped: running full power, using a big antenna. The signal strengthened, and within two (seemingly interminable) minutes, it became clear. Manfred could hear the operator now, but could the operator hear Manfred, in a tent in the middle of a New England forest, with his 100 watts? One hundred watts was more or less the same as a standard incandescent light bulb. But the light bulb's energy had been transformed into radio-frequency electromagnetic waves at 28 million "jiggles per second" as they rushed out through the aluminum antenna in the forest clearing.

Manfred called. There was a slight delay before the Spanish operator said, "Whiskey One Something. Again?"

"Whiskey One Papa. Whiskey One Papa."

This time there was no delay. "Whiskey One Papa, Fifty-Nine URE." (URE is the designation for Unión de Radioaficionados Españoles.)

"Thank you. Five-Nine, Eight." (Manfred's zone, number 8)

"Thanks. Echo Fox Four Hotel Québec—Contest." (The signal was coming from EF4HQ, the headquarters station for the association of Spanish radio amateurs.)

He did it! Manfred had gotten through. And it was a new multiplier, too! The band was open, at least a bit. Now he found other callers. He combed the band up and down—28 MHz was alive. He had to look for every single opportunity while the magic solar energy made it possible.

▷─◠◠◠─╫

You focus now on slaying the dragon, zorching the armored battle tanks. Somehow you survive the hail of sound and fury. You're on your way.

▷─◠◠◠─╫

A second signal popped up from the background hiss—another headquarters station. Things happened faster after that, as signals from the Czech Republic, Slovakia, France, Germany, and Belgium were readable. Now there was Italy, Scotland, Ireland, Morocco, and Poland. *Bang bang bang.*

Gradually, even the loudest voices began to shrink. Finally they were subsumed into the noise, the whiteness, the dull evenness of nothing. It was over. Manfred had tuned his receiver the entire time. It wasn't the time to call CQ. Two-thirds of the stations heard him; the others didn't. The box to the treasure opened up. He was there.

The team was successful in catching the 28 MHz opening: Bouncing between voice and Morse, they nabbed a total of twenty-nine contacts (twenty-one of them on voice). Seventeen of the contacts were valuable new multipliers, most of them from well-equipped headquarters stations.

The final four hours were tough at every site. Each team must have been exhausted, running on fumes after the fuel tank was empty. At the WıP site, "extreme fatigue" hit on Sunday morning. Stefan was affected first, and then it "bit" Manfred an hour later. The men couldn't keep their eyes open, were not able to concentrate, and had trouble pushing the right buttons on the radio and pressing the correct keys on the keyboard. Neither could sit any longer—it was just too uncomfortable—so they operated for a while standing up. Both went outside for a brief bathroom break. It cost them some contacts, but the cool air helped them shake the zombie-like feeling.

In Manfred's own words, "You just had to push yourself through those last hours. Then the machine would work again." The machine, of course, was the human body—stretched to the absolute limit.

On top of the physical and mental limits of human endurance, there had been an equipment failure. Late the previous evening, just as the 3.5-MHz band was opening to Europe, a difficult nighttime path for valuable multipliers, Stefan was unable to contact well-known stations. He could hear them, but none returned his calls. Other WRTC stations were getting through, however. The Germans were falling behind. There seemed to be a one-way street.

Stefan was frustrated. "What should I do?" he asked. Manfred was having trouble thinking clearly, but he crawled underneath the operating table and disconnected the large bandpass filter for the 3.5–3.8-MHz band. Stations began to hear the calls now, yet much critical time had been lost, and the team had missed a number of strong headquarters "beacons" in Europe. The sun was rising on the European end. The window was closing. Closing fast. Stefan managed some American contacts, but the problem had cost them.

⊳⟨⟨⟨⟨⟩⟩⟩⟩⊲

With only ten minutes to go, the race to the finish was in full-bore, flat-out mode. Manfred and Stefan retrieved a burst of energy from somewhere. Yet they never changed their stoic, quiet style. No one could have done it more smoothly—at least it seemed that way to me as I was invited inside the tent, and watched the digital clock on the cluttered operating table move inexorably forward.

Five minutes to go. Now four. Slowly, ever so slowly. Three, two, then one.

"Okay," said Wes. He put his headphones on the desk. "It's over." Both Manfred and Stefan slumped in their chairs and slipped their headphones off. It must have been a relief after such total commitment.

There was little time to celebrate, or even to discuss, what had taken place in the past twenty-four hours. And in only thirty minutes the team would have to wrap up two critical requirements. The first one was to make any post-contest changes to their electronic logbook based on notes they made—either on the computer or with pen and paper—and copy the final contest log onto an external USB drive. The second one was to put the complete recorded audio files, for the entire twenty-four hours for both stations, on the same drive.

That single storage unit, ironically so tiny, neat, and compact, now in the hands of their referee, contained the team's results. Bits and bytes stored in a flash memory the size of a quarter-inch-thick postage stamp represented three years of qualifying, preparation, time and expense, and sometimes nearly superhuman commitment and focus.

It was all over, really over, at 8:30 a.m., July 13, 2014. The team now walked outside. Michael and Derek Bennett, Judy Attaya-Harris, and I gave them a solid round of applause. They seemed a bit overwhelmed. The morning was lovely—fresh and cool.

Large bandpass filters—the biggest of them all would fail late Saturday night and would be bypassed. Photograph: Jim George (N3BB).

*Chapter 19*

# CLOSING IT OUT

~♦~

The top team had a solid lead at 4 a.m. Their margin got even bigger as Dan Craig and Chris Hurlbut built on their early morning results with an additional 407 contacts and thirty-seven multipliers, cruising home with 7.35 million points on the scoreboard. They were the only team to exceed 7 million.

Twenty-five of their new multipliers were the result of the early Sunday opening on 28 MHz, after the band had yielded almost no "usual" daylight-path contacts Saturday. When the band popped open to Europe between five and eight o'clock Sunday morning, a treasure trove of possibilities fell open *if* the teams remembered to look (or to listen). None did it better than the men at K1A, who logged seventy-three contacts in the last few hours, including a whole series of European stations that were weak but were there. Craig and Hurlbut could hear and be heard.

Trailing at a smidgen below 7 million were the second-place Slovakians, who actually had narrowed the gap with the Americans from 530,000 (at 4 a.m.) to 390,000 points at the end of the contest. Overall, the second-place team added forty-two multipliers in the final time segment, along with 466 contacts. Yet it was too late to make a successful final charge all the way to the top.

⊳ ℓℓℓ ╟ⁱⁱⁱ

The judging committee's focus, in terms of the top spots, would be on the bronze medal. The gold and silver were set, barring some extraordinary problems in the data. The next three teams were clustered together, and only with the use of electronic log checking would it be sorted out.

Kevin Stockton and Steve London, who had operated from "Whiskey One Zulu," had been on a tear—a remarkable upward movement—ever since their middle-of-the-pack position after the second hour. Their trajectory was rocket-like: twenty-eighth, thirteenth, sixth, fifth, and now fourth at twenty hours. During the last four-hour stretch, the team was successful, pouncing on the spotty, faint, early morning 28 MHz signals. They made an impressive forty-five contacts there, of which nineteen were new multipliers. Despite having logged the exchange correctly with six previous contacts with the Greek headquarters station, contacts on other bands (on both voice and Morse), the team had somehow, this one time, typed in the geographic zone instead of the headquarters designator. Was it operator fatigue? A short time later, the same sort of error occurred with the German headquarters station, even though they had logged it correctly in eight other prior contacts. The computers later would flag both those errors, but as their clock finally displayed 08:00 Sunday morning, the scoreboard showed them with 4,537 contacts and 420 multipliers, for 6.67 million points. Now they were within about 300,000 points of the Slovakians.

Stockton and London had made nearly 500 contacts in the final five hours, when they were bone-tired and milking the last contacts from both the nighttime bands and the early daylight activity. London, who was stiff and uncomfortable at twenty-one hours, never left his operating seat, even for a bathroom break. "I had to go, but stayed in the chair," he said.

The team at the center of this tight cordon of bronze-medal

contenders included the American Fred Kleber and Canadian John Sluymer, who had operated N1M. Their 400 or so contacts during the last four hours had been coupled with twenty-five additional multipliers. This result was a little off the pace, and they fell 90,000 points behind Stockton and London, even though they had held a healthy 210,000-point margin four hours earlier at 4 a.m.

⊳˷ℓℓℓ˺Ⅲ▸

At the W1P site, Stefan von Baltz took the cell phone from Michael Bennett and looked at the numbers on the small screen. Then he passed the phone over to Manfred Wolf, who at that moment seemed almost too exhausted to take it. It took a while for his eyes to adjust to the tiny font. The two teams at the top had a big lead. In addition, Bennett mentioned that these leaders had been one-two for several hours; they were set. The three-four-five teams were close together, but the final surge by Stockton and London had lifted them into third place, with what seemed to be a safe lead of nearly 100,000 points. Manfred and Stefan apparently were locked in a razor-thin battle with the US–Canadian team, but they were trailing by 30,000 points. Electronic log checking would make the call, but at first glance the bronze medal looked to be out of sight.

For most teams, a fifth-place finish would have been something to treasure. Just to make the top ten in any WRTC would be a badge of honor—your ticket to radiosporting recognition for the rest of your life. But to Manfred Wolf and Stefan von Baltz, it seemed to be just another near miss, another "what could have been." They both looked at the phone. With no spoken words, their disappointment was clear.

It took only an hour to take the station apart. All the equipment was somehow stowed away back into carrying cases. The 4,386 contacts in their electronic log were evidence of what one light bulb's worth of power can do when transformed into electromagnetic energy at the handling of two gifted operators—from a simple installation—in twenty-four hours.

At 9:30 a.m., packed and loaded to the gills once again, the Dodge Charger ferried the team back to the hotel. The trip was anticlimactic. No one said much. Twenty-four hours of serious concentration had ended in a big adrenaline rush at the very end of the contest. It was over.

⊳—⪿⪿⪿—⫞⫾

For the rest of the field, the medals podium would be out of the question. Every team had prepared and come a long (in certain cases, a very long) distance, with lots (a whole lot for some) of equipment, at a considerable expense and commitment of time. Of the fifty-nine teams, it was probably fair to say that twenty of them felt they had a chance to win the WRTC 2014. Of that group, roughly half felt they not only *could*, but *should* win the competition. This field included the top three finishers from Moscow, Brazil, and Finland—the three most recent championships—plus a few new up-and-comers. Make no doubt about it: The WRTC attracted most of the top radiosporting competitors in the world, past and present.

⊳—⪿⪿⪿—⫞⫾

One of the more memorable stories involved the Lithuanian team. They had improved their position to the number-9 slot by 4 a.m., and at the finish had squeaked into sixth place by 3,000 points. All of this, of course, would be run through the software data–based analysis grinder overnight. Yet it was a true tale of grit and determination. Gedas Lucinskas was a veteran contester. He had been a top ten finisher in Moscow. Radiosporting was extremely popular in Lithuania, so his experience and commitment didn't come as a surprise. Yet this sort of comeback was remarkable by any standard. Perhaps this team's dream was still within reach. Perhaps "the third time never lies" after all. Could it be possible for them to rise up with some sort of "accuracy magic" and be a contender for the bronze?

⊳╍ℓℓℓ╍╫╍

Tom Georgens and Dave Mueller turned off their equipment after the final imaginary bell sounded. They were in the number-12 position, just outside their very real target of the top ten. Mueller, who had finished number 4 in Finland and number 2 in Brazil (both with Dan Craig, who now was at K1A), might have been disappointed, but the partnership with Georgens had gone well. They had given it their best shot. Mueller wouldn't have accepted just any offer to compete. All in all, it had been a good weekend. It was over now, and the scoreboard reflected 4,170 contacts and 410 multipliers—5.92 million points. They had just missed 6 million. But the team *had* found the early Sunday opening on 28 MHz and made sixty-two contacts (and nineteen new multipliers) at the end.

Vlad Aksenov and Alexey Mikhailov weren't satisfied with their preliminary number-16 ranking. They had fallen short on contacts (at 3,932) and ended with 5.85 million points, protected by the plywood barrier from the nearby taxiway at Mansfield Airport. They actually had four very solid hours between 4 a.m. and the 8:00 finish. That surge wasn't enough to make a major difference overall. But it did move them ahead of the sponsored Brazilian-Bulgarian Friendship Team, who had taken a very early number-1 position, but then dropped slowly and would finish number 18.

Even though some teams were displeased with how they finished, Sandy Räker and Irina Stieber at N1A were quite satisfied. They felt (self-imposed) pressure to do well, since Räker had qualified—the first woman to do so.

The two women continued to move up through the field to finish at the number-21 spot on the scoreboard, with 3,885 contacts,
395 multipliers, and 5.64 million points. To have finished in the
top half of this group of fierce propagation warriors, none of whom
backed off at all, was an extremely gratifying result. Neither Räker
nor Stieber left her chair the entire twenty-four hours—no potty
break at all. They also found the 28-MHz band opening, in the
form of forty contacts and twenty new multipliers.

Räker commented in more detail: "At the end of [Saturday], I
thought we had missed the [28-MHz] opening because we didn't
have many [contacts] in the log." They actually had twenty-nine
contacts with ten multipliers (on that band). It must not have
seemed like much to Räker at the time, and indeed they were a
bit lower than the middle of the pack. She continued, "I was pretty
concerned and mad at myself that I didn't check [the band] enough.
But during the last hour of the contest the next morning, [28 MHz]
was open and I could contact some of the missing mults. But some
of these stations didn't hear me, while some of the other WRTC
stations got through." The opening must have been spotty and possibly somewhat location dependent. Räker finally managed a very
competitive number of contacts.

She went on to say that the "amazing" 14-MHz band was "open
almost the entire twenty-four hours—very good to Europe." *Wil-
kommen* to New England conditions, Sandy!

The contest was tiring for everyone. Räker elaborated on this:
"At the end of the contest I was all worn out. The last thirty minutes . . . I was very tired and my concentration was bad. But until
then, I wasn't tired at all. Thanks to adrenaline."

Räker later commented at length about the outstanding support she and Stieber received from the site crew: "They were
awesome. They took pictures during the contest, even at night
with our camera. They moved the shades over the tent as the sun
moved around."

A final story captures a personal aspect of this team's WRTC experience. It's far from radiosporting as such; instead, it touches upon our shared humanity. In Sandy Räker's own words: "Another story of our site is the 'rabbit story.' After the contest was over and everything was uninstalled, one of the site crew left with his car, a car that [had been] sitting there for a long time. When he moved his car, a little rabbit family was under there. Dramatically, two of the three baby rabbits died (the tire drove over them). The rabbit mother was gone, and only one little baby rabbit was still sitting there in the sun, first squeaking and then silently watching around. I was sooo sad and couldn't really think any more about the WRTC contest. Our site crew member, Mary, promised to take the animal to a vet so that Tim, Irina, and I could leave the site. Mary updated us on this via email, telling me that the rabbit is okay. They named it 'Radio.'"

(Insert a smiley face here, folks.)

Of course, each team had its own story. One of the most disappointing tales came from Ken Low, who described his K1C effort with Scott Redd as follows:

"After listening to tapes of previous conditions (in the overriding IARU contest) on this date last year, I had dreaded the (low number of contacts) from 2 to 7 a.m. Our strategy was very simple: Focus on (Morse code); run as much as we could; change bands only when needed to keep the rate up; and do enough multiplier hunting to be competitive, but not at the expense of rate. We were successful with everything but the multipliers in the [3.5–3.8 MHz] band, where we were ten multipliers behind the leaders, and the killer 28 MHz multipliers, where a direct European opening happened . . . and we completely missed it. (Sad face) That cost us at least twenty multipliers."

Ken Low continued, "I was absolutely stunned to discover this error after the event. I feel the only explanation for the disaster on the multipliers was the fatigue I felt after twenty-three hours of the most intense contesting experience of my lifetime, coming directly after a long day of setup."

The Low-Redd team had gotten off to a solid start. They were number 8 after the first hour, and number 11 after five hours. But from there they dropped slowly down to the low teens until the band-opening miss, which caused a sudden fade at the end to number 28 on the scoreboard. This was a direct mathematical result of missing those multipliers. Catching that early morning European opening would have put them at least ten spots higher.

Again, to quote Ken Low: "I will be waking up at 3 a.m. thinking about [this] for a very long time."

☞〜℘℘〜┼┉

The Youth Team, with vigor and energy to offset inexperience—at least compared with this elite field—finished a very respectable twenty-ninth place. They *did* hit the band opening and racked up twenty-four multipliers on 28 MHz alone at the end.

Two spots below, at number 31, were the extremely disappointed (and displeased) Estonians at NıT. Their 5.4 million points and 3,776 contacts were far below what a team with their reputation would have expected, nearly a full 1,000 contacts below the leaders. Tonno Vahk's summary comments, which were sent far and wide by email after the competition, made it clear he felt that not only was their site lacking (in his words, it never "got out"), but also the new format implemented by Doug Grant and the WRTC 2014 organizers was flawed. It permitted full-bore two-radio operation, unlike the "lock-out," with its one-transmitter rules favored by the Eastern Europeans and used in the previous WRTC.

This cleavage in terms of contest philosophy appeared to be along East-West lines: "Let 'em rip" versus "careful control." At any rate, nothing seemed to change Vahk's mind that the site limits, real or imagined, multiplied twice by two radios, put them into a hole. He was bitter about it and was not hesitant to say so.

George DeMontrond and John Crovelli ended up at number 45. They were all-in. Neither man took a potty break until Sunday morning. They were successful in finding the 28-MHz band opening, through which they made forty-six contacts and sixteen new multipliers. This little surge helped boost their multiplier total overall to 365, along with 3,679 contacts, for nearly 4.6 million points. DeMontrond had worked tirelessly on the microphone, and they ended up with more than 1,600 voice contacts, or almost 44 percent of the total.

Kondou-san and Hazuki-san at K1O limped home at number 57. They never stopped giving it their best. After all, with $700 of checked luggage and sitting 10,000 miles from home, they were in it to the end. The final scoreboard showed them with 3.76 million points based on 3,305 contacts and 339 multipliers. They also had "hit" the opening to Europe on 28 MHz, with forty-six contacts and sixteen new multipliers. But that was too little, too late. On a positive note, their SkookumLogger program for the Mac never once let them down.

L to R: John Sluymer (VE3EJ), Fred Kleber (K9VV/NP2X), and referee "Tom" Soomets (ES5RY) at their operating position in their tent. Photograph: Bob Wilson (N6TV).

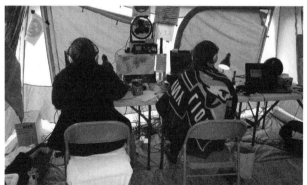

L to R: Irina Stieber (DL8DYL) and Sandy Räker (DL1QQ) operate in the cool morning at N1A. Photograph: Jeffrey Bail (NT1K).

# Chapter 20

# BACK AT HQ

Caravans of exhausted competitors and referees returned to head-quarters. Site managers wrapped up their camping equipment and prepared to go home, even as beam teams returned to the sites and took apart the tents and antenna systems. The network of volunteers had been not only an essential part of the WRTC but an amazing and outstanding component as well.

As part of the financial planning, the standard equipment provided by the WRTC had been sold in advance: towers, antennas, rotators, generators, cabling, tents, and even the latrines. Those modest but efficient communication systems—for portable or fixed residential use—were snapped up by radio amateurs in New England. By the end of Sunday, with a "leave no trace" directive, the sites reverted to their original appearance.

Stefan and Manfred disappeared into the elevator at the hotel, with all their gear in tow on an overloaded trolley. Wes and I remained in the lobby and talked for a bit. He seemed pleased with the apparent top-five finish, but he realized that his team had aimed higher.

A large television screen in the lobby scrolled endlessly through the scores of all fifty-nine teams, so everyone had seen the preliminary order of finish. At this point, the only remaining question involved the bronze medal, and the Americans, Kevin Stockton and Steve London, held a significant advantage over a virtual tie between the fourth- and fifth-place teams. Stockton and London had completed an amazing run-up to challenge for the third spot. It was hard to imagine them not being on the medals podium at the end. I went to my room and decompressed with my notes.

No additional official events were scheduled Sunday, but the World Cup final from Brazil would be shown on television later that afternoon. The bar area would certainly be packed.

The WRTC organizers had arranged for a luncheon, which was spread out on long tables at each end of a large conference room. The room was half-filled when I got there for a serve-yourself spread of cold cuts and soup. The food was tasty and hit the mark. People moved around comparing notes. For me, it was an opportunity to ask several competitors (and referees) about the contest from their perspectives. The talk centered on the standings, of course, yet many other stories of trials and tribulations surfaced.

At one point in the contest, the leader of the Chinese team, "Jack" Zheng, had reached down to get something when his headphones fell off and landed on his computer. The machine died immediately and refused to restart. It took more than an hour for his site crew to locate a suitable replacement.

At K1A, Dan Craig and Chris Hurlbut, who seemed to have had a golden, problem-free, winning competition, actually had to deal with quite a dilemma. When they restarted their special contest software at the beginning of the competition, approximately 200 "practice contacts" they had made prior to the start at 08:00 were sent inadvertently into their computer network. Many of the new stations were displayed as already contacted. They started logging on paper while painstakingly removing each of the false entries one at a time.

Quite a few of the special power supplies for the custom-designed scoreboard computers overheated and failed. These forced the referees at the affected stations to send score updates by text message every fifteen minutes. One referee, a Californian who clearly was an electronics handyman, somehow managed to repair his defective unit in the field while continuing to monitor both of his operators. The unit was put back into service and worked for the rest of the contest.

The remainder of Sunday was, thankfully, unstructured. It was a day of rest and story swapping. The contestants, referees, and other participants (drivers, site managers, other volunteers, video teams, and one writer) were provided downtime for interacting and socializing. Yet for a few—a group essential to the results—the work had just started.

# Chapter 21

# DECISIONS, DECISIONS

Reports began to flow in immediately. Gee-whiz software programs tracked every aspect of radiosporting. Details of every contact were recorded: time, frequency, call sign, and the information sent and received.

Let's say that you were one of the many thousands at home in locations all over the world. You were in a group of individual operators who had operated in the encompassing IARU contest, and of course you had contacted many (or all, as hundreds had focused on) of the WRTC stations. When the competition was over, your wrap-up was simple: You simply needed to run your post-contest summary application, and your data would be compiled and presented in just the right format. You pressed one key. Within seconds, a compilation of your participation in the competition was completed. Another application neatly summarized your score, backed up with the number of contacts, along with multipliers, for every applicable frequency band.

Using an Internet universal resource locator (URL) link for the radiosporting event, you sent in your electronic summary. That was it. A robot somehow appeared and reported back to you if it accepted your formatted data. If not, you were advised as to what needed to be changed.

It was a snap. After ten or fifteen minutes, your experience in the actual contest was over. Now, depending on your location on the planet, it might be time for breakfast, lunch, or supper. Perhaps you were off for a walk in the park, or to bed to make up for a full day—depending on your commitment—of no sleep during the competition. For some of you, the best part of a radiosporting event would be the days following, when battles won and lost would be rehashed and bantered back and forth with friends over lunch, or perhaps over a beer at the pub.

For a tiny group of people, the next twenty-four hours would prove to be anything but routine. For some of them, longtime friendships would be tested.

The WRTC organizing committee had requested that electronic logs be sent in within six hours after the end of the contest. It was important to acknowledge the top finishers and award the medals at the closing ceremony on Monday night. Realistically, the judges had roughly twenty-four hours to sort out the order of finish at the top. Each of the WRTC logs would be processed using a check-off system against the electronic database expanding with input from around the world.

The WRTC itself ended at eight o'clock Sunday morning in New England, which corresponded to Sunday afternoon and evening across Europe. In addition to the American submissions, a significant amount of data came from Europe, where radiosporting was a hotbed of activity. Yet now people had to return to daily life—supper and sleep—and get to work Monday morning.

Logs cascaded in—over 2,500 of them within the first six hours alone. The inflow then continued at a hundred per hour for another six hours. The log-checking process had been developed and refined based on many contests over decades. By now, every conceivable situation that had ever arisen was incorporated into the procedure. Obvious checkpoints included ensuring the accuracy of the information: call sign, time, and the required exchange. In addition, the analysis tested for "cheerleading," which occurred when operators from a home country or region contacted only one favored station (usually from that same area), rather than every operator on an equal basis.

At the end of the contest, the final scoreboard showed Dan Craig and Chris Hurlbut to be nearly 400,000 points ahead of the Slovakian team, who in turn led the third-place team of Kevin Stockton and Steve London by an additional 300,000 points. Those two margins appeared to be insurmountable, and the initial database confirmed that. The gold and silver medals were decided. The data hotshots in the back room ran their regular log-checking procedure and didn't have to do any additional work for the top two positions.

There had been a small spike of German contacts for the women's team of Sandy Räker and Irina Stieber. But that was considered natural, since it was easy to recognize them from their voice operations. They had a nice, respectable finish, but they weren't remotely a factor for the top three spots.

For the next group of finishers, the contenders for the bronze, it was quite another story. The scoreboard indicated that positions three through five were bunched very closely. Even minor adjustments to these scores could make a difference in the order of finish. In addition, the teams in places six through ten were not mathematically out of it. Each of the top ten preliminary scores had tallied at least 6 million points—an amazing achievement. Every one of these teams had racked up over 4,000 contacts from their low-power transmitters and modest antennas. In all but two cases, over 400 different multipliers had been registered. Some serious operating skill was on display here.

By 4 p.m. Sunday, eight hours after the end of the contest

itself, nearly 60 percent of the content of the WRTC logs had been cross-checked. At first, the third- through sixth-place positions were close, yet the number-6 team, the Lithuanians, separated from the top five as the electronic database increased. In any event, Gedas Lucinskas and his partner, Mindis Jukna, had made an astounding comeback.

"Tree" Tyree and Trey Garlough, the software experts, along with the entire judging committee, now focused their attention on the three-four-five grouping.

The scoring of any contest is basically the same: (number of contacts) X (points per contact) X (diversity factor, or multipliers).

The devil is in the details, as always. Different mistakes may reduce a claimed score once the electronic cross-check is completed.

If a contact is disallowed (the term used is "busted") because the call sign is miscopied, then no credit is allowed, and in addition, a one-point penalty is assessed. If only the exchange information is copied incorrectly, it's a simple reduction of that single entry.

If a deleted entry happens to be the only one attached to a multiplier, then that multiplier credit, along with the points for the contact, also is lost.

If an operator takes credit for a contact that is missing from the log of the other station he or she contacted, that is called a "not-in-log," or NIL. This circumstance usually results when the other station's operator actually contacted someone else. The usual cause for this is that two radiosporting operators are very close to each other, near the same frequency. As a result, the mistaken operator not only loses all credit for that contact, but is assessed an equal penalty for any points associated with it.

As each additional batch of logs came in, Tyree and Garlough reran their algorithms. The three teams were so close that a single new log could switch all three positions, especially if a contact with a multiplier attached to it was thrown out. Each of the final contenders for the precious bronze, with the honor of standing on the medals podium at the closing awards ceremony, was treated equally—the process was exactly the same.

Gradually, the incoming data revealed that the (original) fourth-place team on the scoreboard, Fred Kleber and John Sluymer, the operators at N1M, were separated by a widening gap compared with the two remaining candidates for the bronze. Kleber and Sluymer would be number 5, a major distinction, but they wouldn't medal.

More reports continued to come in throughout Sunday. Approximately fifty man-hours of work were put into the analysis. Experts Tyree and Garlough were able to add new logs to their database for fifteen hours, until 11 p.m. Sunday. Along with the judging committee, they had worked continually and were exhausted. A huge amount of quality time had been concentrated on determining the bronze, especially between the Stockton–London team and the Wolf–von Baltz duo, now locked in a virtual tie at numbers three and four.

By this time, the judges and their analytical gurus had obtained over 80 percent of the logs they would ever receive. They had to make a decision. Ever since they started focusing exclusively on the order of the three-four-five teams, Tyree and Garlough had used a technique to look at every contact that did *not* show up in at least ten different WRTC team logs. These contacts tended to be from casual operators around the world. Perhaps these people had just a few minutes to turn on their radios and exchange a report with a handful of competitors active in the contest. This wasn't unusual.

One of these "uniques," as they are called, which was spotlighted by the special software, was the Morse code contact that Steve London had made on 14 MHz with an operator who was located on tiny Lipsi Island, a dot in the Greek Dodecanese within view of the Turkish mainland. Contacts with that operator showed up in no other WRTC team's log, and in only three other logs altogether.

Creating a unique is not bad in itself. It's usually a matter of someone making very few contacts. If the contact is valid, it's counted.

The scores of the final two contenders for the bronze were so close that they literally now were only a handful of contacts apart. In fact, with the scoring algorithm, a single multiplier effectively was equal to ten contacts. The spotlight narrowed on this one contact (fairly or unfairly). It was the only contact between any of the fifty-nine WRTC teams and the operator on Lipsi. (Recall that the Dodecanese Islands counted as a separate "country" from Greece because of the distance from the Greek mainland.)

Three Americans, one Englishman, and a prominent contester from Tajikistan formed the judging committee. By now, they were fully involved with the analysis, working closely with Tyree and Garlough. Several of the group seemed to feel that there was, effectively, a tie. The positions could change with each run of the algorithm once a new handful of the logs was received. Why not call the finish a tie? Award two bronze medals?

But WRTC Chairman Doug Grant had issued a clear mandate: "No ties unless the scores really are exactly the same." Chief Judge Dave Sumner's position was equally clear: "The number is the number." When the exhausted judging group finally ended their session and went to their rooms around midnight Sunday, no one knew who the bronze medalists would be.

Monday morning the judges, along with Tyree and Garlough, reassembled. It still wasn't clear who was number 3 and who was number 4, even after twenty-four hours. There was a "number," all right. But what did it signify? For all purposes, the two scores were essentially, but not exactly, mathematically identical. Grant ordered two versions of the metal scrip for the bronze plaque—one with each team's name on it, along with their call sign—in order to stay prepared for either team to win. But there was only one heavy wooden plaque. Which of those two thin stick-on metallic symbols would go on the plaque for third place?

In addition to the Lipsi contact, the log-checking software had immediately spotted the errors made by Stockton and London during the early morning 28 MHz opening, when they entered the wrong exchange for the Greek and German headquarters stations. Those contacts were removed from the log, and most important, their two multipliers were lost. The Wolf–von Baltz team also had several other contacts removed for logging or copying mistakes, including multiplier credits for South Africa on 7 MHz and Algeria on 21 MHz. No one was perfect. But accuracy would be critical since the two teams were so close.

No electronic log had been received from the operator on the tiny Greek island. That wasn't unusual—he had made only four total contacts and had not sent in his electronic data. Finally, after attempts to contact him had resulted in confusing and conflicting information, and the judges were unable to verify the contact with an acceptable level of confidence—knowing full well the pain it would cause the team losing the bronze—and at the absolute end of the time limit, the committee made its decision. The SV5-Dodecanese contact was nullified, taken out of the log. The points, and more important, the precious multiplier, were lost. That did it. Instead of medaling by a half multiplier, Stockton and London were number 4 by the same half multiplier.

All of this was made especially difficult because the competitors, judges, and referees form an extended worldwide family, with relationships spanning decades in many cases. Yet this is what judges must do in all sports—make the calls and move on.

The Americans and the Germans had been *so* close. It truly was a shame that someone had to be, mathematically at least, number 4, but the final decision was made. Manfred Wolf and Stefan von Baltz would take the bronze.

L to R: Larry "Tree" Tyree (N6TR), Doug Grant (K1DG), and Tim Duffy (K3LR). Photograph: Bob Wilson (N6TV).

"Tree" Tyree (N6TR). Photograph: Bob Wilson (N6TV).

*Chapter 22*

# THE AWARDS

The final events were scheduled for Monday night. A massive dinner would take place in the largest room the hotel had to offer, followed by the awards ceremony and conclusion of the WRTC 2014. Each meal ticket included a table number, and my ticket led me to a table that sat at the extreme rear corner of the room. It was as far from the speakers and awards as possible. Yet any initial disappointment quickly dissipated when my fellow tablemates arrived. They included two key WRTC men and their wives, a Russian WRTC team, and another local couple.

Tom Frenaye, a previous WRTC competitor as well as referee, and his wife sat directly across from me. The ambient crowd noise seemed focused on our rear corner of the room, yet with semi-shouted words, he made it clear that he was pleased with the event. He had served as a member of Doug Grant's organizing committee, with the significant duties of the station locations and site support teams. Frenaye, a civic-oriented man and tactful public figure, previously was the first selectman (mayor) of Suffield, Connecticut. One of Frenaye's key decisions had been adding the man on my right to his team for the event. That man, elbow to elbow with me, was Dean Straw. (How could I have been so lucky?) He and his wife now lived in San Francisco; however, he had worked for the ARRL in Connecticut for many years. At the league, he became known as the "antenna expert" who had overseen and added to the *ARRL Antenna Book* since he took on the position of editor in 1993.

During Straw's fifteen-year tenure at the ARRL, he also served as editor for numerous technical publications in the areas of antennas and transmission lines and was a frequent contributor to the ARRL's monthly magazine, *QST*. A Google search of his name and "N6BV," his current call sign, reveals the technical contributions the highly talented Straw has made to the hobby of amateur radio: There are hundreds of links to impressive technology publications, the historic notations of his childhood as a novice radio ham in Hawaii, and the talks and presentations he has made over the years at radio clubs and conferences (as well as glowing tributes to him from people who have built antennas or have solved problems following his guidelines).

Spending time with Straw was a study in both his humility and understatement. When I asked him about the operating sites, he simply answered that he thought the site selections were as fair as possible in terms of signal propagation. They weren't identical, but he believed that Rich Assarabowski and the site selection people had found a number of excellent locations. Each of them had been analyzed, and many had been tested in actual practice, with their signals—as measured with receivers in Europe—"falling into a reasonably narrow range of equality." He never said the exact words, but it was clearly my impression that Straw believed any difference in scores would be determined primarily by the competitors' strategies and their operating capabilities.

The WRTC itself was unique. The specific rules and officiating made that clear. To deal with this form of radiosporting, one that required separate communication sites—usually in remote locations without regular power or facilities—created special demands. To make those separate sites equal, to level the playing field, demanded nearly superhuman levels of coordination and planning. The Russians probably came the closest to perfection in the Moscow WRTC (in spite of a historic heat wave, punctuated by a dangerous thunderstorm in the middle of the competition), where the tent sites were clustered in a few closely bunched groupings on flat land.

All in all, the WRTC 2014 organizers had planned very well. Everyone agreed that the event was supported by an impressive volunteer network. Yet the claims of certain location advantages, or disadvantages, continued to persist.

On my left was Anatoly Polevik, who in turn sat next to his WRTC teammate, Mikhail "Mike" Klokov. Both men resided in Novosibirsk, the third most populous city in Russia. Located in Siberia, three time zones ahead of Moscow and six ahead of GMT, Novosibirsk is to Moscow as New York is to Los Angeles in terms of time difference.

Polevik was a real estate developer and, according to his webpage, the champion motorcycle racer in Russia at one point. He wrote that he was "a born contest man." That seemed clear from the descriptions of his huge home amateur radio station. He traveled widely, citing stops on all continents with the exception of Antarctica.

Polevik had brought his wife and children with him, and they planned to continue touring the East Coast down to Florida and then visit California before returning home. He spoke relatively fluent English, with a to-the-point manner. It was clear from his responses to my questions about his WRTC experience that he considered his operating site far from optimum. He and Mike had been at site 9A, an eastern grouping of six teams and two spare sites. It was one of two clusters closest to the Atlantic Ocean, along with the Myles Standish State Forest set of seventeen operating sites.

The Russian team on my left expressed disappointment at what they considered to be a suboptimal location. To my right, the renowned developer of the High Frequency Terrain Analysis

(HFTA) software described attempts to level the playing field as much as humanly possible. It was a fascinating dinner discussion.

As the desserts were being served, Doug Grant stepped up to the microphone. He was pleased that the competition and all tours had concluded safely, and thanked everyone for traveling to Boston to take part. He acknowledged the terrific support of hundreds of volunteers.

Grant then gave recognition to the media publicity. He specifically mentioned that the book you're now reading would be available, along with a video from James Brooks, himself a world-class contest operator and several-time WRTC competitor. Brooks lived in Singapore, held the call sign 9V1YC, and ran a television postproduction company responsible for editing, coloring, and packaging programs for such well-known channels as National Geographic, the History Channel, and the Food Network. In addition, he had produced documentary-style videos of the most recent WRTC competitions, as well as dozens of "DX-peditions," some perilous, to unusual and hard-to-reach places. For the WRTC 2014, Brooks had worked with Chuck Green of Perpetual Motion Media, a New England videographer, who had taken many of the shots at the hotel and in the field at the operating sites.

Next, Grant commented on the Ham Widows' Ball, and his remarks brought a huge cheer from the very ham widows who had found it such a hit. Bill Vinci, who sat with his partner, Rusty Epps, took a poorly disguised bow amid the tumult. "Oh, and there was also a contest," said Grant with a Cheshire-cat look.

Grant went on to acknowledge and thank each man on his organizing committee, as well as the fifty-one qualifying teams, the defending champions, the Youth Team, the two Wild Card teams, and the four sponsored duos. "It was a fair competition," he concluded.

Then he introduced Rui Amen, the director of tourism for the Azores. Amen's agency had sponsored one of the four teams that provided significant financial backing for the competition. The team consisted of two of the well-known "Flying Finns," Martti Laine and Ville Hiilesmaa. The duo had competed as "Radio Team Azores." They had had a bit of a rough go, given their previous track records, finishing at number 47 from their W1R station.

Amen narrated a travelogue-style video about the Azores, which are one of the two autonomous regions of Portugal (along with the Madeira Islands), located almost 840 miles nearly due west from the mainland. These "nine wonderful islands," per Amen, had beautiful weather (mid-50s to mid-80s). No recent eruptions had taken place, thank goodness, from the volcanic peaks, which were some of the tallest mountains on Earth when measured from their seafloor bases.

Next, Grant introduced Dave Sumner, who would present the medals along with several special awards. Sumner began by thanking his four committee members. Then he acknowledged Larry Tyree and Trey Garlough, the data experts and radio contest analysis mavens.

"There were 3,400 logs submitted to the database inputs," Sumner went on. "Sixty percent of the contest contacts have been cross-checked." Sumner pointed out that the best hour of the twenty-four-hour duration had been the 10 p.m. local hour, during which the fifty-nine teams made a mind-numbing 15,573 contacts. That averaged 115 per hour for each individual operator! Indeed, radiosporting was a contact sport.

He continued by saying that the WRTC stations were "very comparable in signal strength," and referenced specific comparisons done in Dubai and Arizona during the contest. "It was a level playing field," he concluded.

Awards were presented for the highest number of voice contacts (with at least 35 percent of the total made on Morse). The winners, team "Kilo-One-Mexico," or K1M, included two well-known, voice-centric Italians: Carlo de Mari, IK1HJS, and Fabio Schettino, I4UFH. They were pleased, even with their fifty-eighth-place finish; they had focused on voice. (George DeMontrond and John Crovelli might have given them a run for their money, but they went for a higher overall point total.)

The award for the highest number of Morse code contacts (with a minimum of 35 percent on voice) went to the forty-fourth-place team of Luxembourger Philippe Lutty, LX2A, and Romanian Andy Ruse, YO3JR, at the N1S station. The highest-multiplier award went to K1A, where Dan Craig (N6MJ) and Chris Hurlbut (KL9A) had dominated the competition nearly throughout. "You'll see these guys again in a minute," said Sumner. Obviously, there was no suspense regarding the overall winner.

Things were getting really interesting now. The medals awards would follow. Yet, a prized recognition, the award for highest accuracy, was next up. "The most accurate log, with a total score reduction of only one percent," said Sumner, who paused for dramatic effect, "goes to W1P." Loud cheers erupted, and Sumner waited to continue. "DJ5MW and DL1IAO." Accuracy is next to godliness among radiosporting cognoscenti.

I smiled as Manfred Wolf and Stefan von Baltz bounded up to the awards podium and received the plaque. "Now don't leave. Stay where you are," said Sumner, and with that he launched into an explanation of the excruciating efforts that the judging committee had gone to. "We looked at the three-four-five teams so carefully. It was difficult. Finally, the decision came down to eight thousand points between the third- and fourth-place teams. That's six-tenths of one multiplier. Or, said another way, one minute out of twenty-four hours on the air!" Sumner closed with an acknowledgment that the W1Z team was close, so very close. But they ended as number 4.

With that, he turned to Manfred Wolf and Stefan von Baltz and announced that "Whiskey One Papa" had won the bronze medal. The room erupted again in celebration—well, most of the room. The combination of the accuracy award plus the bronze capped off a wild ride for the two Germans.

Following that, the silver medal was presented to the Slovakians, Rastislav Hrnko, OM3BH, and Jozef Lang, OM3GI, at the W1L site. The gold-medal award followed, for Dan Craig and Chris Hurlbut at K1A. These two teams had broken away so distinctly that there was little uncertainty surrounding the final two places. Yet they had made great accomplishments and deserved significant honors. As a bonus for winning the gold, Craig and Hurlbut won a week's vacation, expenses paid, to those "wonderful islands," the Azores.

The three medalist teams went to the podium. The American national anthem was played for the winning team, and the evening neared its conclusion.

Following the anthem, Tine Brajnik, retired brigadier general in the Slovenian army, and an elite-level contester who had competed with his teammate (they had finished 11th), declared the WRTC 2014 competition closed. Brajnik, head of the WRTC Sanctioning Committee, requested proposals for the WRTC 2018. After a week of well-planned activities, good weather, and spirited on-air radiosporting competition, the formal part of the championship was over.

The final scores of the top five teams were:

1. **GOLD**: Dan Craig and Chris Hurlbut
   (both USA) K1A: N6MJ and KL9A; 7,184,844

2. **SILVER**: Rastislav Hrnko and Jozef Lang
   (both Slovakia) W1L: OM3BH and OM3GI; 6,816,144

3. **BRONZE**: Manfred Wolf and Stefan von Baltz
   (both Germany) W1P: DJ5MW and DL1IAO; 6,421,383

4. Kevin Stockton and Steve London
   (both USA) W1Z: N5DX and N2IC; 6,413,056

5. Fred Kleber (USA) and John Sluymer
   (Canada) N1M: K9VV and VE3EJ; 6,302,958

Anatoly Polevik (RC9O, center) with family and friend Tom Berson (ND2T, on the right). Photograph: Bob Wilson (N6TV).

Chief Judge Dave Sumner (K1ZZ) at the awards ceremony. Photograph: Bob Wilson (N6TV).

Award for
WRTC winners.
Photograph: Bob
Wilson (N6TV)

Doug Grant
awards the Azores
Islands trip prize
to a delighted
winning team of
Dan Craig and
Chris Hurlbut.
Photograph: Bob
Wilson (N6TV).

# Chapter 23

# ONE HAPPY TABLE

In most major award ceremonies, the after-parties are a key por-tion of the celebration. Apparently, the WRTC was yet not ready for an Oscar-like series of soirées. *Vanity Fair* did not reserve a suite at the nearest Four Seasons or at the Ritz. There was no lim-ousine awaiting the winners to whisk them away. Yet the atmo-sphere truly was celebratory. Of course the winners, Americans Craig and Hurlbut, were one center of attention. They had been expected to excel, and indeed they had—in dominating fashion. The team of Hrnko and Lang had a similar cluster of well-wishers, and clearly had done a fantastic job in the competition. One area of serious "making whoopee" was centered at the Germans' table, where not only Manfred Wolf and Stefan von Baltz celebrated, but Sandy Räker and Irina Stieber, along with referees Wes Kosinski and Rusty Epps, congregated. Judy Attaya-Harris had come for the awards banquet; her outgoing personality blended in perfectly with the more serious Teutonic crowd.

The room was still buzzing with excitement long after Tine Brajnik's official declaration of closure. It was an international cel-ebration. It was good.

L to R: Manfred Wolf (DJ5MW), referee Wes Kosinski (SP4Z), and Stefan von Baltz (DL1IAO) with the Most Accurate plaques. Photograph: Michael Hoeding (DL6MHW).

L to R: Irina Stieber, Judy Attaya-Harris, and Sandy Räker at the awards ceremony. Photograph: Jim George (N3BB).

# Chapter 24

# A LOOK BACK

The decision on the third-place team, significant because of the bronze medals and the podium appearance, had been extremely difficult. It was gut wrenching. And, in retrospect, it was flawed. That specific contact with Lipsi Island actually *had been made*. The audio recording confirmed it, as did a later communication by the operator on Lipsi, though there had been some uncertainty at the time.

History is more nuanced, however. For example, the Stockton–London team had been successful in finding that dramatic and unpredictable atmospheric opening early Sunday morning on 28 MHz. However, when the operator (who no doubt was tired with lack of sleep and the constant pressure) had made those forty-five contacts and nineteen precious new multipliers during the last four hours, two of them had been typed into the log incorrectly. The dominant German headquarters station was logged with their geographic zone, not with their official DARC (Deutscher Amateur Radio Club) acronym. Likewise the Greek HQ station was entered with its zone, and not the Greek IARU abbreviation. So, even with elite operators, among the most highly skilled in the world, mental fatigue and the fog of competition had an effect. If either one of those contacts had been logged accurately, along with their potential new multipliers, the results would have shifted between the third and fourth positions.

☞〰〰╫⊪

The WRTC 2014 was officially over. By nearly all conventional standards, it was a major success. The tours toured. The excursions excursed (don't look it up—the word doesn't exist). No one was lost or injured. Sixty-five operating sites had been located in a densely populated area of the United States, a region with varied terrain. Permission to access public and private grounds had been secured, with stipulations as unusual as the one that any unexploded ordnance must be left alone! Fifty-nine complete, field-ready operating communications positions were put into place, including a unique and effective antenna that was designed and manufactured in Arkansas, plus a technique that allowed each tower-antenna-rotator system to be raised and later lowered with a small team of people. Site equivalence had been tested by stations operating in the IARU contest the previous year, a test that included actual signal-strength measurements taken from Europe, Asia, and the Americas. On top of that, using detailed topographical data, the takeoff angles of the transmitted signals at each site were simulated with sophisticated computer analyses. Every site was approved on the basis that it was within a range of signal strength equality, neither too high nor too low.

Of course, some sites *were* a bit better and some a bit worse. There was no such thing as being absolutely equal. The intent was to put the sites into a "nondeterminative" category. The results were to be based on operating skills: knowledge of band conditions, the ability to read different accents of human voices, Morse code capability, strategies for when to "run" and when to "search and pounce," and overall determination and stamina.

A significant majority agreed that the championship had met

its goals. Everyone agreed that the top teams were outstanding. However, a vocal minority left the competition convinced that their site had been a limitation, a significant restriction to their radio-sporting potential.

Following the contest, email was distributed showing terrain propagation simulations, based on the High Frequency Terrain Analysis software. These indicated a substantial differentiation in takeoff angles, differences that would either optimize or diminish gain into Europe. One specific program focused on the Estonians' location, since Tonno Vahk believed that they "didn't get out" as well as others. Compared with the first- and second-place teams, the Estonians' site was equal for higher-angle takeoff conditions but perhaps was at a disadvantage for the extent of four to six degrees into Europe, an important range of conditions.

An interesting and different examination was published in the *National Contest Journal* (*NCJ*) issue of September–October 2014. In this analysis, the final scores of all fifty-nine teams were plotted against the figure of merit (FOM) that had been developed for each of the sites by Rich Assarabowski, based on Dean Straw's HFTA software. The results showed a broad scatter plot with *no correlation* of these scores to the FOMs. A "zero" figure-of-merit corresponded to completely flat ground. All but three sites had an overall positive rating, and 0.5 was the broad average for all fifty-nine sites. Overall, there were "no obvious terrain advantages" cited by the author.

In a related method, the scores of the top teams were plotted against their site FOMs:

1. Craig/Hurlbut      +0.4

2. Hrnko/Lang      +1.3

3. Wolf/von Baltz      +0.2

4. Stockton/London      +0.5

5. Kleber/Sluymer      +1.6

As a matter of interest, the Estonians' site rated as a +0.9, while the defending Russians' location was rated at a –0.3, one of only three sites with a slightly negative FOM.

The Moscow WRTC 2010 has been cited as a very high standard for site equality. This put the focus on operator ability as the primary determinant—as it should have been. The close-together clusters of operator tents and antennas were on flat ground. Aside from the unusual weather (clearly out of the organizers' control), the major concern would have been interference from the strong signals sent out by neighboring teams.

An analysis published in the November–December 2014 *NCJ* compared the *range* of contacts and scores between Moscow 2010 and New England 2014. Perhaps surprisingly, the high-low scores in Moscow were 4.1–1.7 million, for a 2.37:1 spread, while the comparable results in New England were 7.2–3.4 million, for a 2.12:1 range. The number of contacts also was more tightly centered in New England, with a high-low range of 4,572–3,107 (1.47:1), compared with 3,440–1,973 (1.74:1) in Moscow. These indicate that the results in 2014 were more closely grouped.

A different take on the results was published more recently, in the January–February 2015 issue of the *NCJ*. A fairly strong correlation was shown between the final results at the WRTC 2014, when compared with the so-called "pile-up" contests that were offered on an optional basis prior to the actual competition in the field. Prerecorded audio tracks featured various radio call signs, simulated with different speeds (in Morse). Appropriate interference and a few offsetting signals were thrown in for good measure. The best results were associated with those operators who generally posted the highest scores for the WRTC 2014.

The author wrote, "I concluded that a very important trait for a top contester is to get a caller's complete call sign correct the first time. This maintains a constant rate that could result in more contacts during the contest." He added, "There appears to be a general trend. The better your ability to pick out call signs in the (Morse) pile-up competition, the better you performed overall in the WRTC 2014."

Nate Moreschi, who summarized the IARU contest in a recent *QST* article, put his experiences (as an WRTC competitor at the K1R site) this way: "It really is like drinking radiosport through a fire hose." It is that, indeed.

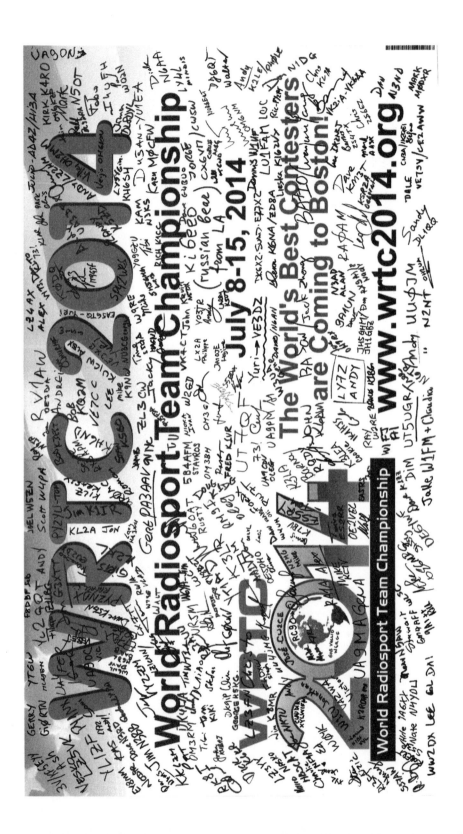

*Appendix*

# THE RESULTS OF THE WORLD RADIOSPORT TEAM CHAMPIONSHIP

## July 20–21, 1990
## Seattle, Washington, United States

| Ranking | WRTC call sign | Operators' call signs | | Final score |
|---|---|---|---|---|
| 1 | N7MJZ/WG | K1AR | K1DG | 263.35 |
| 2 | W7WKR/WG | K7JA | W9RE | 255.39 |
| 3 | W7TSQ/WG | KQ2M | KR0Y | 254.30 |
| 4 | K7ZR/WG | VE7CC | VE7SV | 247.44 |
| 5 | W7NG/WG | DL5XX | DJ6QT | 247.11 |
| 6 | N7NKG/WG | LZ1MS | LZ2PO | 244.86 |
| 7 | K7SS/WG | G3YDV | G4BUO | 239.21 |
| 8 | KS7L/WG | EA5BRA | EA9EO | 237.56 |
| 9 | W7KJJ/WG | UA9AM | UW9AR | 236.45 |
| 10 | K7LXC/WG | AA4NC | W7EJ | 235.12 |
| 11 | K7WA/WG | UA1DZ | RB5IM | 232.78 |
| 12 | N7ZZ/WG | I2UIY | IK2DVG | 229.06 |
| 13 | KE7V/WG | OH1XX | OH8PF | 228.06 |
| 14 | W7KT/WG | OK1RI | OK2FD | 225.38 |
| 15 | N7AY/WG | YU1RL | YT3AA | 220.55 |
| 16 | K7RI/WG | HA6NY | HA0MM | 219.85 |
| 17 | K7LR/WG | UW3AA | UA9SA | 209.91 |
| 18 | KR7G/WG | UW0CA | UW0CN | 209.59 |
| 19 | K7LZJ/WG | JE1CKA | JE1JKL | 195.64 |
| 20 | W7FR/WG | PY4OD | PY5EG | 193.06 |
| 21 | N7TT/WG | FD1NYQ | F2CW | 177.43 |
| 22 | KM7E/WG | JJ3UHS | JM3JOW | 172.45 |

## July 13-14, 1996
## San Francisco Bay Area, California, United States

| Ranking | WRTC call sign | Operators' call signs | | Final score |
|---|---|---|---|---|
| 1 | W6X | KR0Y | K1TO | 761,829 |
| 2 | K6T | K4BAI | KM9P | 678,132 |
| 3 | W6R | K6LL | N2IC | 655,720 |
| 4 | K6P | VE3EJ | VE3IY | 647,112 |
| 5 | K6C | K4UEE | N6IG | 644,059 |
| 6 | W6T | K5ZD | WX3N | 616,308 |
| 7 | W6D | K1KI | K3UA | 606,550 |
| 8 | W6Q | 9A3A | S53R | 598,272 |
| 9 | W6V | KF3P | KR2J | 577,575 |
| 10 | W6P | K8CC | K5GO | 568,435 |
| 11 | K6V | W2GD | W0UA | 568,378 |
| 12 | K6W | N6TV | K7SS | 556,928 |
| 13 | W6I | K1AR | K1DG | 547,404 |
| 14 | W6Y | DL1IAO | DK3GI | 545,756 |
| 15 | K6D | DL5XX | DL1VJ | 532,728 |
| 16 | K6R | LZ1SA | LZ2PO | 531,552 |
| 17 | K6G | NP4Z | WC4E | 527,592 |
| 18 | W6A | K3LR | WA8YVR | 523,672 |
| 19 | K6X | UA3DPX | RZ9UA | 518,666 |
| 20 | K6Z | JH4NMT | JE3MAS | 512,535 |
| 21 | W6S | LY2IJ | LY1DS | 509,392 |
| 22 | W6B | S59A | S56A | 507,318 |
| 23 | K6Y | OK1CF | OK2PAY | 499,796 |
| 24 | W6H | RW1AC | RV1AW | 497,965 |

| | | | | |
|---|---|---|---|---|
| 25 | K6I | JH7PKU | JO1BMV | 488,940 |
| 26 | K6S | ON4UN | ON9CIB | 480,326 |
| 27 | W6U | EA1AK | EA4KR | 470,744 |
| 28 | W6G | JE1JKL | JH7WKQ | 470,237 |
| 29 | K6U | SM3DMP | SM3CER | 465,075 |
| 30 | W6O | ZS6EZ | ZS6NW | 461,553 |
| 31 | K6O | WN4KKN | N6TR | 454,476 |
| 32 | W6E | EA7TL | EA9KB | 445,356 |
| 33 | K6N | YT1AD | YU1RL | 437,167 |
| 34 | W6W | LU6ETB | LU/OH0XX | 437,016 |
| 35 | K6J | N2NT | KZ2S | 426,656 |
| 36 | W6K | F6FGZ | F5MUX | 418,375 |
| 37 | K6A | JH4RHF | JA8RWU | 412,388 |
| 38 | K6H | DJ6QT | DJ2YA | 411,376 |
| 39 | K6K | UT5UGR | UT4UZ | 398,399 |
| 40 | K6F | IT9BLB | IT9VDQ | 385,280 |
| 41 | K6M | GI0NWG | G3OZF | 383,437 |
| 42 | K6B | 9A9A | 9A3GW | 383,166 |
| 43 | K6Q | VE7NTT | VE7CC | 362,440 |
| 44 | K6E | HA0MM | HA0DU | 357,885 |
| 45 | W6Z | VK5GN | VK2AYD | 343,604 |
| 46 | W6J | SP6AZT | SP9FKQ | 330,876 |
| 47 | W6L | UN4L | UN2L | 309,518 |
| 48 | K6L | SP9HWN | SP9IJU | 298,178 |
| 49 | W6N | I4UFH | I2VXJ | 269,028 |
| 50 | W6M | PY0FF | PY5CC | 231,066 |
| 51 | W6C | IN3QBR | IT9TQH | 185,070 |
| | | | | |
| | W6F | OH2IW | OH1JT | 565,000[a] |
| | AH3C[b] | YL2KL | YL3DW | |
| | AH3D[b] | BA1OK | BA4RC | |

[a]*Approximate score—damaged log file.*   [b]*Demonstration team.*

## July 8-9, 2000
## Bled, Slovenia

| Ranking | WRTC call sign | Operators' call signs | | Final score |
|---|---|---|---|---|
| 1 | S584M | K1TO | N5TJ | 965.31 |
| 2 | S587N | RA3AUU | RV1AW | 910.86 |
| 3 | S582A | K1DG | K1AR | 867.15 |
| 4 | S517W | DL1IAO | DL2MEH | 866.10 |
| 5 | S537L | OH1EH | OH1NOA | 846.15 |
| 6 | S511E | DL6FBL | DL1MFL | 845.19 |
| 7 | S523W | UT4UZ | RW1AC | 837.19 |
| 8 | S573O | 9A9A | 9A3GW | 825.02 |
| 9 | S519I | KQ2M | W7WA | 820.29 |
| 10 | S533G | DL6RAI | OE2VEL | 813.16 |
| 11 | S581I | VE7ZO | VE3EJ | 812.11 |
| 12 | S518N | K6LA | K5ZD | 808.71 |
| 13 | S531R | K1ZM | N2NT | 804.89 |
| 14 | S524G | LY1DS | LY4AA | 793.93 |
| 15 | S512T | LY3BA | LY2BM | 789.31 |
| 16 | S548X | UT5UGR | UU2JZ | 782.03 |
| 17 | S549L | RZ9UA | UA3DPX | 780.90 |
| 18 | S536P | HA3OV | HA3NU | 770.73 |
| 19 | S539D | ON4WW | ON6TT | 762.44 |
| 20 | S562P | IK2QEI | I2VXJ | 759.55 |
| 21 | S567F | EA3NY | EA3KU | 755.26 |
| 22 | S528D | OM3BH | OM3GI | 753.65 |
| 23 | S526O | K8NZ | W2GD | 751.33 |
| 24 | S568Y | G3SXW | G4BUO | 745.19 |

| 25 | S544Z | YT1AD | YU7NU | 741.77 |
|----|-------|-------|-------|--------|
| 26 | S577V | UA9BA | RN9AO | 738.10 |
| 27 | S546Q | K4UEE | N6IG | 733.57 |
| 28 | S522R | LW9EUJ | LU7DW | 726.77 |
| 29 | S574V | K9TM | N2IC | 719.80 |
| 30 | S542B | 9A3A | 9A2AJ | 714.54 |
| 31 | S583D | DL2CC | DL5XL | 712.67 |
| 32 | S588S | WC4E | W0UA | 709.69 |
| 33 | S572L | ZS6EZ | ZS4TX | 705.67 |
| 34 | S534J | K4BAI | K6LL | 703.51 |
| 35 | S529A | 5B4WN | 5B4LP | 697.96 |
| 36 | S541F | S59A | S58A | 694.05 |
| 37 | S571W | K3NA | N6TV | 691.31 |
| 38 | S532N | PP5JR | PY2NY | 689.28 |
| 39 | S521H | VE7SV | VA7RR | 683.68 |
| 40 | S586U | OK1QM | OL5Y | 679.75 |
| 41 | S514U | JM1CAX | JO1RUR | 667.35 |
| 42 | S566Z | K9ZO | K7BV | 661.86 |
| 43 | S578R | PY5CC | PY1KN | 653.71 |
| 44 | S538F | S50U | S51TA | 644.92 |
| 45 | S561C | VE3BMV | VE3KZ | 644.16 |
| 46 | S543C | F6BEE | F6FGZ | 642.02 |
| 47 | S547B | SP8NR | SP9HWN | 638.69 |
| 48 | S527K | JH4NMT | JK3GAD | 618.51 |
| 49 | S513A | JA8RWU | JH4RHF | 617.99 |
| 50 | S516M | EA7GTF | EA7KW | 582.68 |
| 51 | S563X | N3AD | N3BB | 567.29 |
| 52 | S564Q | VK4EMM | VK4XY | 511.92 |
| 53 | S576K | I5NSR | I5JHW | 431.76 |

## July 13–14, 2002
## Helsinki, Finland

| Ranking | WRTC call sign | Operators' call signs | | Final score |
|---|---|---|---|---|
| 1 | OJ3A | N5TJ | K1TO | 1,629,798 |
| 2 | OJ8E | RA3AUU | RV1AW | 1,619,226 |
| 3 | OJ2V | DL2CC | DL6FBL | 1,608,673 |
| 4 | OJ3R | N6MJ | N2NL | 1,560,008 |
| 5 | OJ8K | KQ2M | W7WA | 1,479,470 |
| 6 | OJ5A | VE3EJ | VE7ZO | 1,473,127 |
| 7 | OJ1M | K5ZD | K1KI | 1,469,255 |
| 8 | OJ6E | UT4UZ | UT3UA | 1,468,064 |
| 9 | OJ5W | LY1DS | LY2TA | 1,459,744 |
| 10 | OJ5M | DK3GI | DL1IAO | 1,456,840 |
| 11 | OJ6W | OE2VEL | OE9MON | 1,436,448 |
| 12 | OJ6C | RW1AC | RW3QC | 1,414,100 |
| 13 | OJ5U | N6RT | N2NT | 1,412,640 |
| 14 | OJ8W | 9A9A | 9A5E | 1,405,837 |
| 15 | OJ7M | SP3RBR | SP8NR | 1,402,440 |
| 16 | OJ2F | N6TJ | N6AA | 1,391,088 |
| 17 | OJ3T | RZ9UA | UA9MA | 1,390,795 |
| 18 | OJ2H | N5RZ | K2UA | 1,388,670 |
| 19 | OJ8A | K1AR | K1DG | 1,382,400 |
| 20 | OJ2J | HA1AG | HA3OV | 1,368,432 |
| 21 | OJ3N | N2IC | K6LL | 1,355,940 |
| 22 | OJ4M | K3LR | N9RV | 1,347,612 |
| 23 | OJ3D | W4AN | K4BAI | 1,347,107 |
| 24 | OJ2Y | UA2FZ | RW4WR | 1,331,623 |

| 25 | OJ4N | ON6TT | ON4WW | 1,301,248 |
|----|------|-------|-------|-----------|
| 26 | OJ2Q | YU7BW | YU1ZZ | 1,300,734 |
| 27 | OJ6X | OH1MDR | OH1MM | 1,293,414 |
| 28 | OJ7C | ES5MC | ES2RR | 1,288,254 |
| 29 | OJ2Z | G4PIQ | G4BWP | 1,277,950 |
| 30 | OJ6N | OK2FD | OK2ZU | 1,274,577 |
| 31 | OJ1S | SP7GIQ | SP2FAX | 1,234,317 |
| 32 | OJ5T | SM5IMO | SM3SGP | 1,214,742 |
| 33 | OJ7X | S50A | S59AA | 1,210,147 |
| 34 | OJ4S | JM1CAX | JE1JKL | 1,205,008 |
| 35 | OJ7N | YL2KL | YL3DW | 1,196,424 |
| 36 | OJ3X | 5B4ADA | 5B4WN | 1,186,950 |
| 37 | OJ7S | N5KO | N1YC | 1,142,882 |
| 38 | OJ1X | K1ZM | N6ZZ | 1,139,230 |
| 39 | OJ5E | OH6EI | OH2XX | 1,131,630 |
| 40 | OJ1F | NT1N | AG9A | 1,105,645 |
| 41 | OJ5Z | F6FGZ | F5NLY | 1,086,750 |
| 42 | OJ8N | YT1AD | YU7NU | 1,069,820 |
| 43 | OJ7W | UA9BA | RN9AO | 1,052,480 |
| 44 | OJ6K | VE7SV | VE7AHA | 1,045,980 |
| 45 | OJ4A | DJ6QT | DL2OBF | 1,005,259 |
| 46 | OJ1C | LU7DW | LU1FAM | 986,930 |
| 47 | OJ7A | PP5JR | PY1KN | 978,021 |
| 48 | OJ1N | EA3AIR | EA3KU | 954,380 |
| 49 | OJ8L | S56M | S57AL | 883,545 |
| 50 | OJ1W | ZS6EZ | ZS4TX | 880,065 |
| 51 | OJ6Y | IK2QEI | I4UFH | 878,349 |
| 52 | OJ4W | UN9LW | UN7LAN | 699,732 |

## July 8-9, 2006
## Florianópolis, Brazil

| Ranking | WRTC call sign | Operators' call signs | | Final score |
|---------|----------------|-----------------------|------|-------------|
| 1 | PT5M | VE3EJ | VE7ZO | 2,439,380 |
| 2 | PW5C | N6MJ | N2NL | 2,317,456 |
| 3 | PT5Y | K1DG | N2NT | 2,098,060 |
| 4 | PW5X | UT4UZ | UT5UGR | 2,024,496 |
| 5 | PT5D | IK2QEI | IK2JUB | 1,987,080 |
| 6 | PT5P | DL6FBL | DL2CC | 1,978,320 |
| 7 | PT5N | 9A8A | 9A5K | 1,962,177 |
| 8 | PW5Q | N0AX | KL9A | 1,958,928 |
| 9 | PT5R | RW3QC | RW3GU | 1,945,174 |
| 10 | PT5Q | W2SC | K5ZD | 1,944,320 |
| 11 | PT5L | YT6A | YT6T | 1,937,647 |
| 12 | PT5K | KH6ND | N6AA | 1,907,788 |
| 13 | PT5W | LY2TA | LY2CY | 1,871,793 |
| 14 | PW5W | RA3AUU | RV1AW | 1,845,432 |
| 15 | PT5B | OH2UA | OH4JFN | 1,804,495 |
| 16 | PW5B | SP7GIQ | SP2FAX | 1,769,625 |
| 17 | PT5I | YL2KL | YL1ZF | 1,747,392 |
| 18 | PW5U | XE1KK | XE1NTT | 1,698,200 |
| 19 | PT5X | PY2NY | PY2EMC | 1,683,825 |
| 20 | PW5K | ES5TV | ES2RR | 1,659,948 |
| 21 | PT5E | K1LZ | LZ2HM | 1,504,464 |
| 22 | PW5Y | K4BAI | KU8E | 1,492,416 |
| 23 | PT5U | K5TR | KM3T | 1,489,911 |
| 24 | PW5V | RW4WR | UA9CDV | 1,457,868 |

| 25 | PW5Z | YO9GZU | YO3JR | 1,441,440 |
| 26 | PW5I | ZS4TX | N2IC | 1,431,848 |
| 27 | PW5O | S50A | S59AA | 1,404,920 |
| 28 | PW5G | IZ3EYZ | 9A1UN | 1,362,812 |
| 29 | PW5L | LZ4AX | LZ3FN | 1,359,579 |
| 30 | PW5D | K1ZM | K1KI | 1,349,969 |
| 31 | PT5J | N6BV | AG9A | 1,333,789 |
| 32 | PW5F | F6BEE | W2GD | 1,308,496 |
| 33 | PT5V | 9A6XX | DJ1YFK | 1,295,728 |
| 34 | PT5G | N9RV | K3LR | 1,212,120 |
| 35 | PW5A | LU1FAM | LU5DX | 1,168,500 |
| 36 | PW5M | PY2NDX | UU4JMG | 1,147,722 |
| 37 | PT5C | OH1JT | OH2IW | 1,136,600 |
| 38 | PT5A | 5B4WN | 5B4AFM | 1,128,519 |
| 39 | PW5P | OZ1AA | SM0W | 1,098,880 |
| 40 | PT5O | HP1WW | N5ZO | 1,095,276 |
| 41 | PT5T | PY2YU | PY1NX | 1,055,240 |
| 42 | PT5F | RA3CO | RW3FO | 960,690 |
| 43 | PW5T | UA9AM | RZ3AA | 938,685 |
| 44 | PW5J | P43E | WA1S | 878,712 |
| 45 | PW5N | JK2VOC | JA2BNN | 842,289 |
| 46 | PW5E | BA4RF | BA7NQ | 534,744 |

## July 10–11, 2010
## Moscow, Russia

| Ranking | WRTC call sign | Operators' call signs | | Final score |
|---|---|---|---|---|
| 1 | R32F | RW1AC | RA1AIP | 4,098,162 |
| 2 | R33A | ES5TV | ES2RR | 4,084,889 |
| 3 | R33M | N6MJ | KL9A | 3,942,904 |
| 4 | R39D | S50A | S57AW | 3,907,540 |
| 5 | R34P | K5ZD | W2SC | 3,889,908 |
| 6 | R32K | RV3BA | RA3CO | 3,776,544 |
| 7 | R32R | LY9A | LY6A | 3,615,024 |
| 8 | R31X | UA3DPX | UA4FER | 3,594,820 |
| 9 | R37M | G4PIQ | G4BUO | 3,558,636 |
| 10 | R36C | LY9Y | LY7Z | 3,502,044 |
| 11 | R33L | VE3DZ | VE3XB | 3,494,064 |
| 12 | R38F | UA9AM | RU9WX | 3,472,950 |
| 13 | R33G | N2NT | K3LR | 3,445,825 |
| 14 | R31U | UU4JMG | UR0MC | 3,417,154 |
| 15 | R34O | HA3OV | HA6PX | 3,389,750 |
| 16 | R36Y | OH2UA | OH4JFN | 3,348,636 |
| 17 | R34W | OM3BH | OM3GI | 3,316,098 |
| 18 | R39M | N2IC | N6TV | 3,306,000 |
| 19 | R32C | DL6FBL | DL3DXX | 3,259,720 |
| 20 | R37L | YO3JR | YO9GZU | 3,241,690 |
| 21 | R37Q | 5B4WN | 5B4AFM | 3,176,899 |
| 22 | R34C | VE3EJ | VE7ZO | 3,156,659 |
| 23 | R36O | RX9TL | RL3FT | 3,132,600 |
| 24 | R38O | UN9LW | UN7LZ | 3,059,836 |

| 25 | R31A | K6XX | N6XI | 3,059,256 |
|----|------|------|------|-----------|
| 26 | R36F | UA9CLB | UA9CDV | 3,056,132 |
| 27 | R38K | N5DX | K5GO | 3,051,171 |
| 28 | R38X | RW6HX | RW6HA | 3,043,110 |
| 29 | R31D | N4TZ | N5AW | 2,917,434 |
| 30 | R34D | UA9ONJ | RO9O | 2,893,623 |
| 31 | R32Z | K1ZM | K1LZ | 2,883,624 |
| 32 | R32O | YV1DIG | W2GD | 2,881,142 |
| 33 | R37A | W4PA | K6LA | 2,863,900 |
| 34 | R32W | IK2QEI | IK2NCJ | 2,850,850 |
| 35 | R31N | VE7CC | VE7SV | 2,836,990 |
| 36 | R36Z | OH6UM | OH7JT | 2,826,820 |
| 37 | R38N | OE3DIA | OE6MBG | 2,782,195 |
| 38 | R36K | 4O3A | 4O7NT | 2,729,580 |
| 39 | R38W | OM2VL | OM3RM | 2,460,692 |
| 40 | R37P | I2WIJ | IK1HJS | 2,452,533 |
| 41 | R39A | YT1AD | YT3W | 2,363,631 |
| 42 | R37U | PY8AZT | PY2NDX | 2,292,269 |
| 43 | R34X | RA3DOX | RV3FM | 2,149,552 |
| 44 | R39R | VK2IA | VK6LW | 2,024,358 |
| 45 | R34Z | 9K2RR | 9K2HN | 1,982,231 |
| 46 | R33U | JK3GAD | JH4RHF | 1,951,796 |
| 47 | R36W | F6BEE | F5JSD | 1,743,714 |
| 48 | R33Q | EA8CAC | EA8DP | 1,729,761 |

## July 12–13, 2014
## New England, United States

| Ranking | WRTC call sign | Operators' call signs | | Final score |
|---|---|---|---|---|
| 1 | K1A | N6MJ | KL9A | 7,184,844 |
| 2 | W1L | OM3BH | OM3GI | 6,816,144 |
| 3 | W1P | DJ5MW | DL1IAO | 6,421,383 |
| 4 | W1Z | N5DX | N2IC | 6,413,056 |
| 5 | N1M | K9VV | VE3EJ | 6,302,958 |
| 6 | W1A | LY9A | LY4L | 6,129,420 |
| 7 | W1D | K1LZ | YT6W | 6,126,504 |
| 8 | N1K | DK6XZ | DK9IP | 6,096,060 |
| 9 | K1D | UR0MC | VE3DZ | 6,064,890 |
| 10 | K1V | G0CKV | M0DXR | 6,008,327 |
| 11 | K1L | S50A | S57AW | 5,935,524 |
| 12 | K1S | W2SC | N2NL | 5,923,270 |
| 13 | K1K | RL3FT | RA3CO | 5,896,797 |
| 14 | W1S | F8DBF | F1AKK | 5,882,778 |
| 15 | W1M | 4O3A | HA1AG | 5,874,264 |
| 16 | N1F | RW1A | RA1A | 5,850,670 |
| 17 | N1G | RX3APM | RV1AW | 5,840,310 |
| 18 | N1Z | PY1NX | LZ3YY | 5,776,323 |
| 19 | N1R | UA3DPX | UA4FER | 5,677,604 |
| 20 | K1R | N4YDU | N3KS | 5,637,042 |
| 21 | N1A | DL1QQ | DL8DYL | 5,635,070 |
| 22 | K1N | OE3DIA | E77DX | 5,605,488 |
| 23 | K1I | UU4JMG | UU0JM | 5,582,475 |
| 24 | W1W | OH2UA | OH6KZP | 5,568,180 |

| 25 | W1C | 9A5K | 9A1TT | 5,543,250 |
|----|-----|------|-------|-----------|
| 26 | K1G | 9A6XX | 9A1UN | 5,541,416 |
| 27 | K1P | M0CFW | G1ORTN | 5,533,399 |
| 28 | K1C | KE3X | K0DQ | 5,512,230 |
| 29 | K1T | IZ1LBG | WQ2N | 5,491,056 |
| 30 | N1O | RC9O | UA9PM | 5,463,552 |
| 31 | N1T | ES5TV | ES2RR | 5,401,340 |
| 32 | W1U | LZ4AX | LZ3FN | 5,377,152 |
| 33 | K1Z | VE7CC | VE7SV | 5,331,680 |
| 34 | K1W | K6AM | N6AN | 5,267,133 |
| 35 | N1V | K7RL | KL2A | 5,199,210 |
| 36 | N1U | K8MR | K9NW | 5,137,945 |
| 37 | N1L | KU1CW | EA5GTQ | 5,118,876 |
| 38 | N1W | PY2YU | PY2NDX | 5,073,630 |
| 39 | W1F | CT1ILT | CT1BOH | 5,031,380 |
| 40 | W1O | OM2VL | OM3RM | 4,949,288 |
| 41 | W1N | 5B4WN | 5B4AFM | 4,834,860 |
| 42 | W1V | R9DX | UA9CDV | 4,820,352 |
| 43 | W1G | F4DXW | F8CMF | 4,783,096 |
| 44 | N1S | LX2A | YO3JR | 4,711,516 |
| 45 | N1D | NR5M | W2GD | 4,581,480 |
| 46 | W1T | AD4Z | W4UH | 4,532,602 |
| 47 | W1R | OH2BH | OH2MM | 4,500,684 |
| 48 | W1B | OE2VEL | OE5OHO | 4,488,405 |
| 49 | N1C | IK2NCJ | IK2QEI | 4,473,720 |
| 50 | N1P | CX6VM | LU1FAM | 4,418,156 |
| 51 | N1B | YL1ZF | YL2GQT | 4,375,566 |
| 52 | W1I | W2RE | WW2DX | 4,293,500 |
| 53 | N1N | KH6ND | KH6SH | 4,258,656 |
| 54 | K1B | W9RE | N5OT | 4,220,138 |
| 55 | K1F | VY2ZM | KK6ZM | 4,212,430 |
| 56 | W1K | BA5CW | BA7IO | 4,088,500 |
| 57 | K1O | JH5GHM | JA1OJE | 3,759,171 |
| 58 | K1M | IK1HJS | I4UFH | 3,512,224 |
| 59 | K1U | KF5EYY | YT1AD | 3,382,155 |

# INDEX

NOTE: Page numbers in *italics* indicate a photograph.

# F

# G

# H

# I

# J

# ABOUT THE AUTHOR

J. K. (Jim) George (N3BB) has been an avid amateur radio enthusiast, with special love for Morse code and radiosporting, since his
teenage years—over fifty years ago. He started his hobby listening to shortwave broadcasts and long-distance AM radio stations.
He is a graduate of Virginia Tech, where he was president of the
VT Amateur Radio Club and was inducted into the Virginia Tech
Academy of Engineering after serving as Chair of the Advisory
Boards for both Electrical and Computer Engineering, as well as
the College of Engineering. He is an avid Virginia Tech sports fan.
He completed his graduate school at Arizona State University, spe-
cializing in semiconductor materials physics.

Jim's professional career was primarily at Motorola, where he
worked in the semiconductor business for nearly forty years and
served as corporate vice president for nearly fifteen years. Following his retirement, Jim was a founding member and served for
several years on the advisory board for KUT-FM, the public radio
station of the University of Texas at Austin. Jim enjoys reading and
writing and has been a member of a small men's book club in Austin for over a decade. He and his wife live in the Hill Country west
of Austin, and have three grown children and five grandchildren.

His first book was *Reunion*, a novel about a difficult relationship between a father and son, as well as life-long friendships within a unique high-school peer group.